Contents

Number Patterns and Relationships

Week 1	Exploring Patterns	2
Week 2	Patterns and Relationships	12
Week 3	Patterns and Graphs	22
Week 4	Variables and Equality	32
Week 1	Practice	42
Week 2	Practice	43
Week 3	Practice	44
Week 4	Practice	45

Week 1 — Exploring Patterns

Lesson 1

Key Idea

Same-step patterns are patterns in which the amount of change or growth is the same from one set to the next.

Ask the following questions to help you look for a pattern.

- What is changing from one set to the next?
- What stays the same from one set to the next?
- How would you create the next set?

Try This

Describe what changes in each pattern. Sketch and label the next figure in the pattern.

❶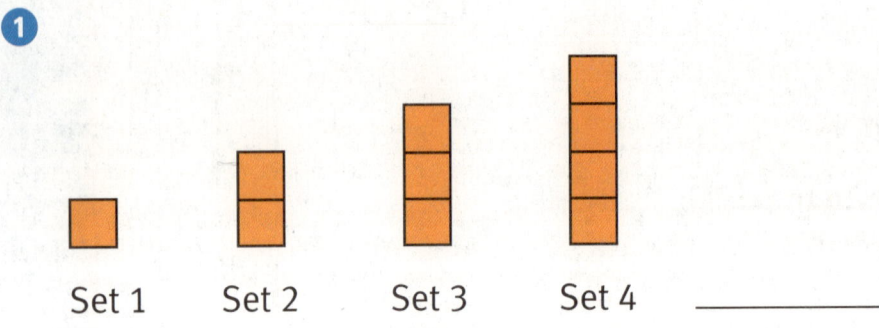

Set 1   Set 2   Set 3   Set 4   _____

_____

_____

❷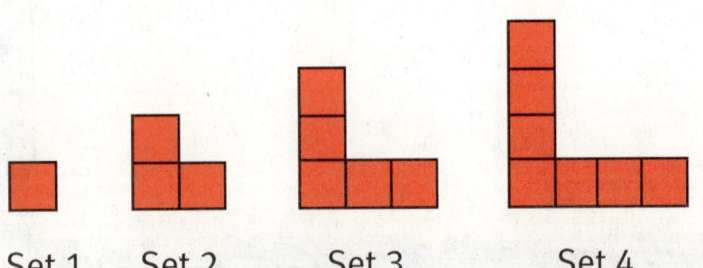

Set 1   Set 2   Set 3   Set 4   _____

_____

_____

2 Number Patterns and Relationships • Week 1

Practice
Sketch and label the next two terms in each pattern.

3

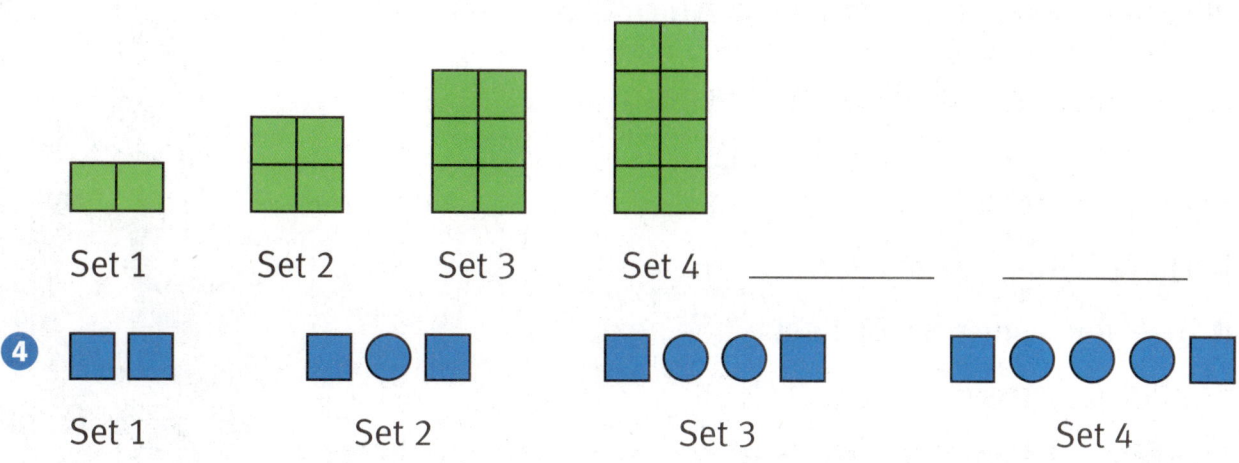

Set 1   Set 2   Set 3   Set 4   _____   _____

4   ▪▪   ▪○▪   ▪○○▪   ▪○○○▪

Set 1   Set 2   Set 3   Set 4

_____   _____

Reflect
Mandy had 5 pennies in her piggy bank on Monday. Each day she put 10 more pennies into the bank. How much money will she have in the piggy bank on Friday? Explain.

Monday	Tuesday	Wednesday	Thursday	Friday
5¢	15¢	25¢	35¢	____

_____
_____
_____
_____

Exploring Patterns • Lesson 1

Week 1 — Exploring Patterns

Lesson 2

Key Idea

Changing-step patterns are patterns in which the amount of change or growth changes in a regular and predictable way from one set to the next.

Ask the following questions to help you identify a changing-step pattern:

- What changes from one set to the next?
- What stays the same from one set to the next?
- How would you create the next set?

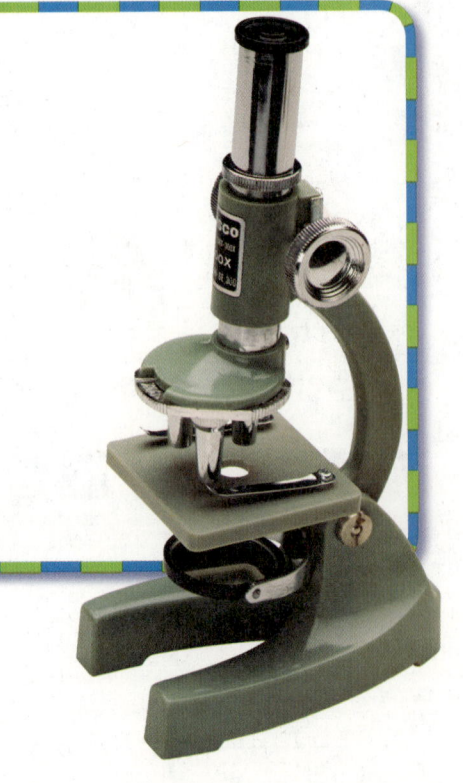

Try This

Describe the changes in each pattern. Then sketch and label the next set in the pattern.

1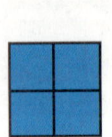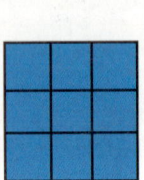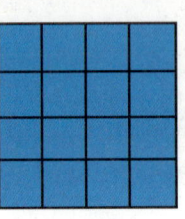

   Set 1   Set 2   Set 3   Set 4   _____

_____

_____

2

   Set 1   Set 2   Set 3   _____

_____

_____

_____

Practice
Sketch and label the next term in the pattern.

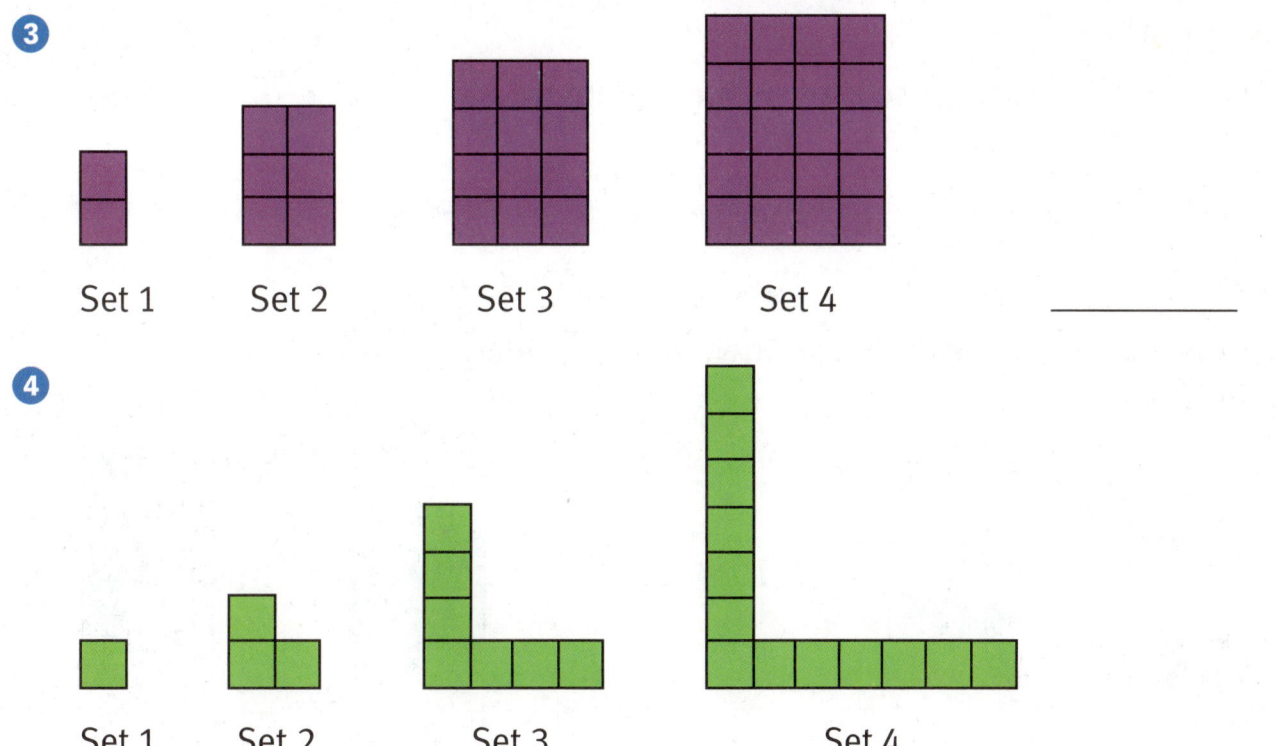

Reflect
Look for a pattern in the table. If the pattern continues, how many chin-ups will Jarrod do on Friday? Explain your answer.

Jarrod's Chin-up Chart				
Monday	Tuesday	Wednesday	Thursday	Friday
8 chin-ups	10 chin-ups	13 chin-ups	17 chin-ups	

_____
_____
_____

Week 1 Exploring Patterns

Lesson 3

Key Idea
You can use clues from the sets given in a pattern to identify missing sets.

Try This
Look for the changes in each pattern. Then draw and label the missing set.

1 Set 1, Set 2, _____, Set 4, Set 5

2 Set 1, Set 2, _____, Set 4

3 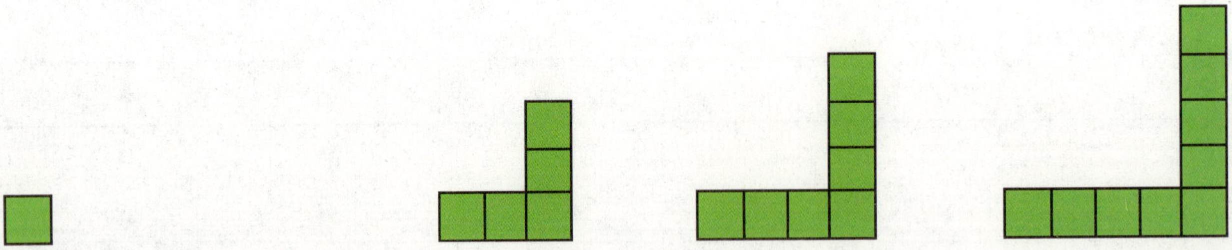 Set 1, _____, Set 3, Set 4, Set 5

6 Number Patterns and Relationships • Week 1

Practice

Look for the changes in each pattern. Tell whether it is a same-step growing pattern or a changing-step growing pattern. Then draw and label the missing set.

④ Set 1    _____   Set 3 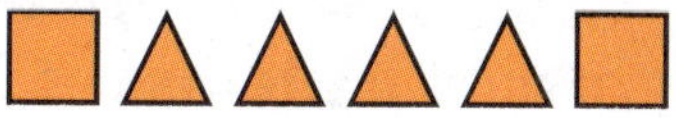   Set 4

⑤ Set 1    Set 2    Set 3    _____   Set 5

Reflect

Do you think it is easier to find a missing set in a same-step or a changing-step growing pattern? Explain.

_____
_____
_____
_____

Exploring Patterns • Lesson 3

Week 1 — Exploring Patterns

Lesson 4

Key Idea
You can use pattern blocks to create your own growing patterns.

Try This

Draw and label the next set for each pattern. Use the pattern to answer the questions.

1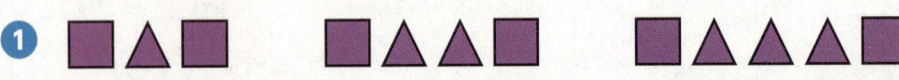

Set 1   Set 2   Set 3   _____

a. Is this a same-step or a changing-step growing pattern?

_____

b. How is the pattern changing from set to set?

_____

2

Set 1   Set 2   Set 3   Set 4   _____

a. Is this a same-step or a changing-step growing pattern?

_____

b. How is the pattern changing from set to set?

_____

_____

8 Number Patterns and Relationships • Week 1

Practice

Draw your own pattern in the space below. Exchange your pattern with a partner, and have your partner answer each question.

❸ Is this a same-step or a changing-step growing pattern?

_____

❹ How is the pattern changing from set to set?

_____

❺ What would the next set of the pattern look like? Describe it.

_____

Reflect

Use different shapes to create the first few sets of a pattern. Explain how the pattern changes from set to set.

Exploring Patterns • Lesson 4

Week 1 Exploring Patterns

Lesson 5 Review

This week you explored patterns. You examined same-step patterns and changing-step patterns.

Lesson 1 Sketch and label the next set in the pattern.

1

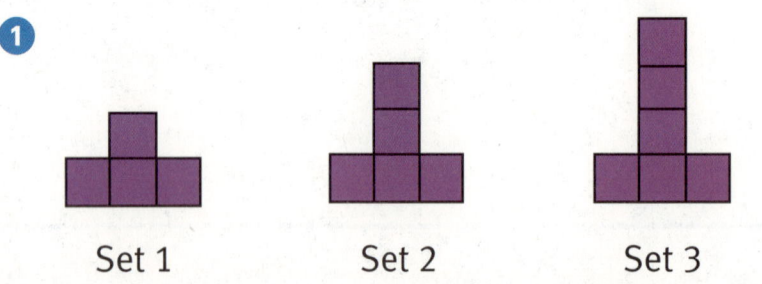

Set 1   Set 2   Set 3   _____

2

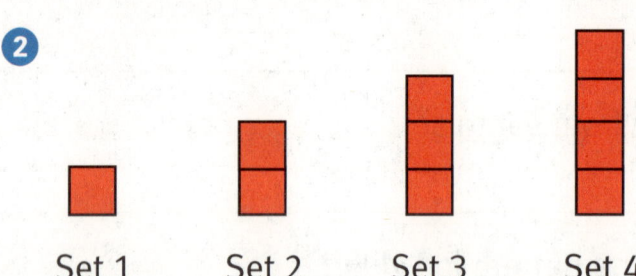

Set 1   Set 2   Set 3   Set 4   _____

Lesson 2 Sketch and label the next set in the pattern.

3

Set 1   Set 2   Set 3   _____

Reflect
Tell whether each pattern above is a same-step or a changing-step growing pattern.

_____
_____
_____

10 Number Patterns and Relationships • Week 1

Lesson 3 Sketch and label the missing set in the pattern.

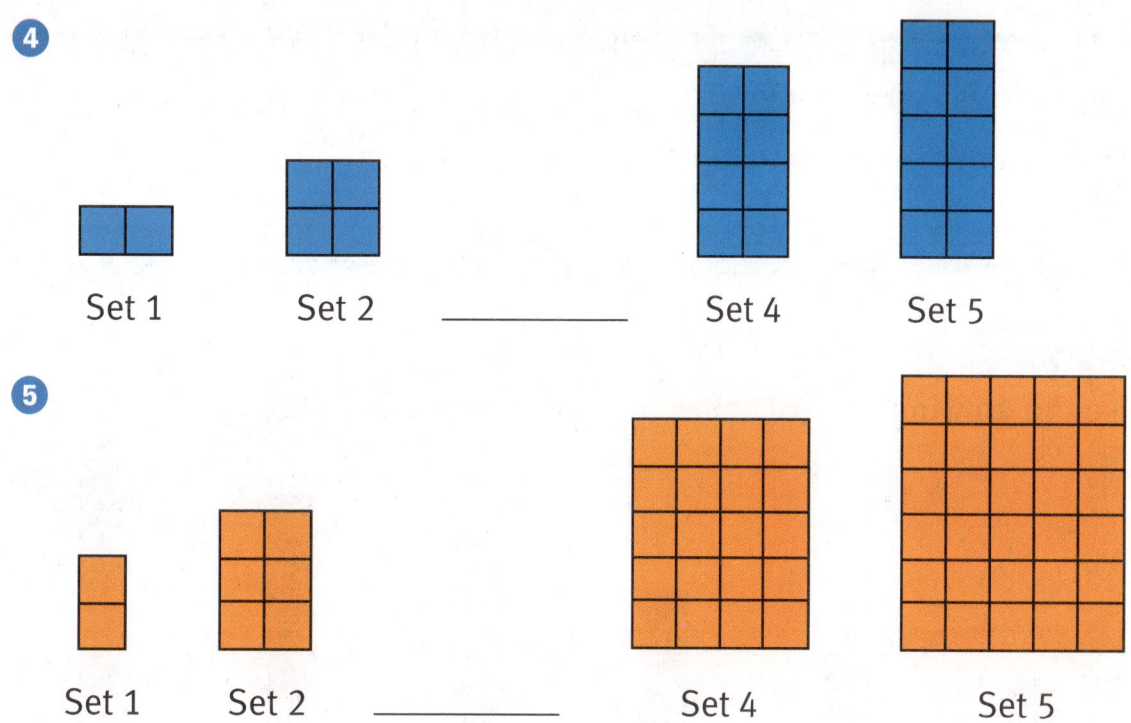

④ Set 1 Set 2 _____ Set 4 Set 5

⑤ Set 1 Set 2 _____ Set 4 Set 5

Lesson 4

⑥ Use pattern blocks to design Set 1 of a pattern. Then show Sets 2 and 3.

Reflect

Describe how to find the missing set. Then draw and label the missing set.

Set 1   Set 2   _____   Set 4

_____

_____

_____

_____

Week 2 — Patterns and Relationships

Lesson 1

> **Key Idea**
> You can use numbers to represent the growth in patterns.

Try This
Use the growing pattern below to answer each question.

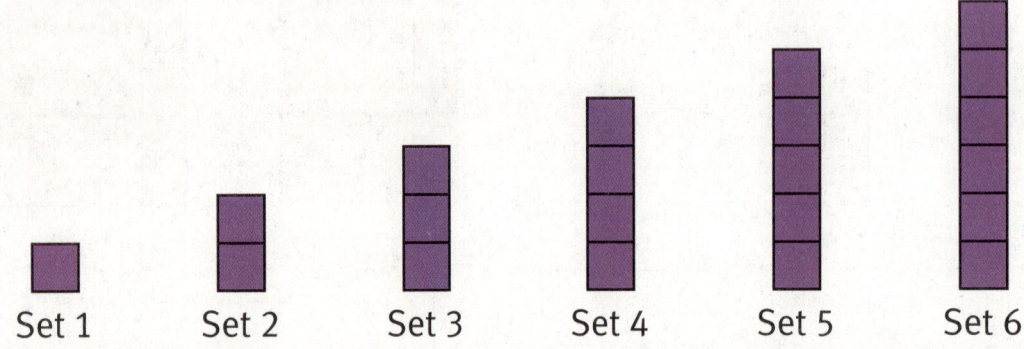

1. Use the pattern to complete the table below.

Set	Number of Squares
1	
2	
3	
4	
5	
6	

2. What patterns do you notice in the table between the set number and the number of squares?

_____

_____

3. Use the pattern from Problem 2 to predict the number of squares in the tenth set without building or drawing all the sets up to the tenth.

_____

12 Number Patterns and Relationships • Week 2

Practice
Use the pattern to complete the table.

4

Set 1   Set 2

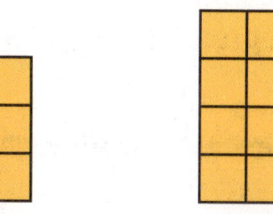
Set 3

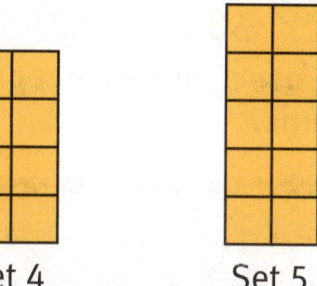
Set 4   Set 5

Set	Number of Squares
1	
2	
3	
4	
5	

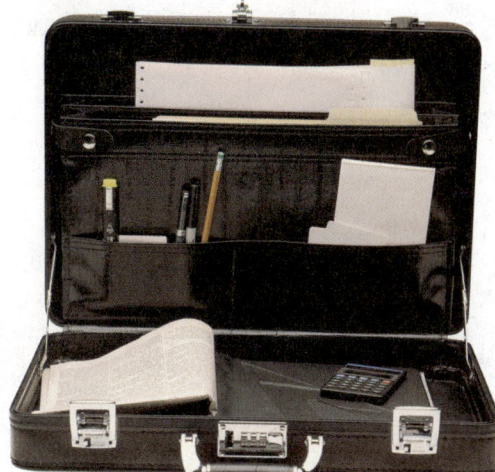

5. What patterns do you notice in the table above between the set number and the number of squares?

_____

6. Use the pattern from Problem 5 to predict the number of squares in the tenth set without building or drawing all of the sets up to the tenth.

_____

Reflect
When you use a pattern table to help you create a pattern with shapes, what is the smallest number of sets you need to figure out the pattern? Explain.

_____
_____
_____

Patterns and Relationships • Lesson 1

Week 2
Patterns and Relationships
Lesson 2

Key Idea
You can use numbers to represent the growth in patterns.

Try This
Use the growing pattern below to answer each question.

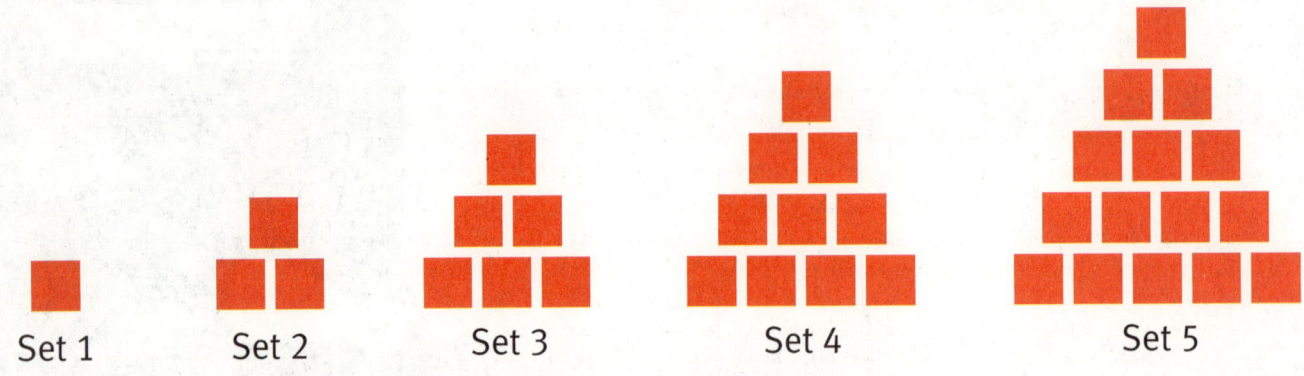

Set 1   Set 2   Set 3   Set 4   Set 5

1 Is this a same-step or a changing-step growth pattern?
_____

2 Describe the growth shown in the pattern.
_____

3 How many squares were used to create each set?
_____

4 Use your answers to complete the table below.

Set	Number of Squares
1	
2	
3	
4	
5	

14 Number Patterns and Relationships • Week 2

Practice
Use the pattern to complete the table.

5

  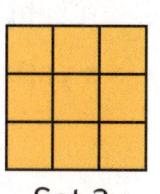  

Set 1   Set 2   Set 3   Set 4   Set 5

Set	Number of Squares
1	
2	
3	
4	
5	

6 What patterns do you notice in the table above between the set number and the number of squares?

_____

7 Use the pattern to predict the number of squares in the tenth set without building or drawing all of the sets up to the tenth.

_____

Reflect
What do you notice about the table for a changing-step pattern? How does it compare with the table for a same-step pattern?

_____

_____

_____

Patterns and Relationships • Lesson 2

Week 2 — Patterns and Relationships

Lesson 3

> **Key Idea**
>
> The pattern tables are examples of input/output tables.
>
> For each input value (the set number), there is a certain output value (number of squares).
>
> Input/output tables can also be used to model real-world situations.

Try This

Thomas mows lawns in his neighborhood to earn money. He earns $8 for each lawn he mows. Use the input/output table to answer each question.

Input (lawns mowed)	Output (money earned)
1	$8
2	$16
3	$24
4	$32
5	$40

1. How much money would Thomas earn if he mowed 6 lawns? Explain how you found your answer.

   _____

   _____

2. Write a mathematical rule for determining the amount of money Thomas will earn for any given number of lawns mowed.

   _____

Practice
Complete the table and answer each question.

3 Movie tickets cost $6 each. Complete the input/output table.

Input (number of tickets)	Output (total cost)
1	$6
2	
3	
4	
5	

4 How much would it cost to buy 4 movie tickets?

_____

5 Write a mathematical rule for determining the total cost of tickets for any given number of tickets.

_____

_____

6 For every hour Lisa drives, she uses 2 gallons of gasoline. Her gas tank holds 18 gallons when it is full. Complete the input/output table.

Input (hours of driving)	Output (gas remaining in her tank)
1	16 gallons
2	14 gallons
3	
4	
5	

7 How much gasoline is in Lisa's tank after 5 hours of driving?

_____

8 Write a mathematical rule for determining the amount of gas remaining for any given number of hours driven.

_____

_____

Reflect
What is different about the lawn mowing input/output table and the gasoline input/output table?

_____

_____

_____

Patterns and Relationships • Lesson 3   **17**

Week 2 — Patterns and Relationships

Lesson 4

> **Key Idea**
> You can use input/output tables to help you make choices.

Try This

Anna's neighbors have hired her to pet sit their dog for seven days. They have offered two different options for being paid.

- **Option 1:** Anna receives $10 for the first day and an additional $2 per day after the first day.

- **Option 2:** Anna receives $1 for the first day. Every day after the first day she receives an additional amount that is $1 more than the previous day.

1. Complete the input/output tables for each option.

Option 1		Option 2	
Day	Total Amount Earned	Day	Total Amount Earned
1	$10	1	$1
2	$12	2	$3
3		3	$6
4		4	
5		5	
6		6	
7		7	

2. If Anna chooses Option 1, how much money will she be paid? If Anna chooses Option 2, how much money will she be paid?

18 Number Patterns and Relationships • Week 2

Practice

Jim was hired to do yard work for his neighbor. The neighbor expects the work to last 5 days, but it could last 7 days. Payment options are as follows:

- **Option 1:** Jim receives $12 for the first day and $2 per day after the first day.

- **Option 2:** Jim receives $2 for the first two days. Every day thereafter he receives an amount that is $1 more than the previous day.

Complete the tables to find Jim's total earnings for the week for each option.

3 Complete the table for each option.

Option 1		
Day	Amount Earned for the Day	Total Earnings
1	$12	
2		
3		
4		
5		
6		
7		

Option 2		
Day	Amount Earned for the Day	Total Earnings
1	$2	
2		
3		
4		
5		
6		
7		

4 How much will Jim earn for 7 days if he chooses Option 1? _____

5 How much will Jim earn for 7 days if he chooses Option 2? _____

6 If the job is for only 5 days, which option will pay better? _____

Reflect

What kind of growth pattern is shown by Option 1? What kind of growth pattern is represented by Option 2?

Patterns and Relationships • Lesson 4

Week 2 — Patterns and Relationships

Lesson 5 Review

This week you explored patterns and relationships. You looked at how visual patterns can be related to number patterns. You also learned about input/output tables and solving problems.

Lessons 1 and 2

Complete the table for the pattern shown below.

1

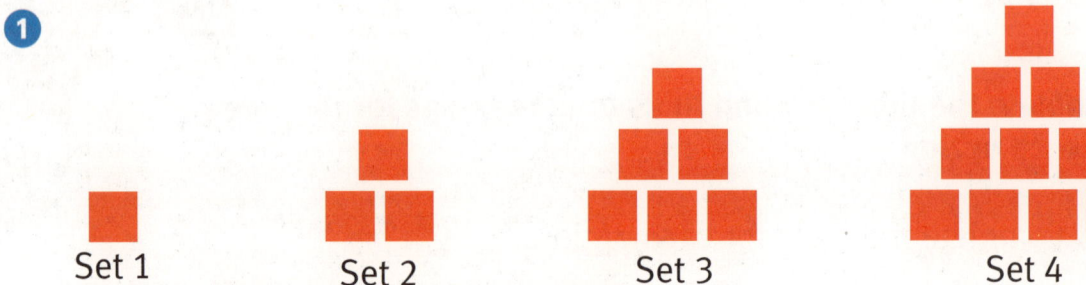

Set 1   Set 2   Set 3   Set 4

Set	Number of Cubes
1	
2	
3	
4	

Reflect
How is the pattern changing?

_____

20   Number Patterns and Relationships • Week 2

Lesson 3 Complete the input/output table.

② The bookstore sells pencils for 15¢ each.

Input (number of pencils)	Output (total cost)
1	15¢
2	
3	
4	

Lesson 4 A bathtub holds 60 gallons of water. When the drain plug is pulled, 12 gallons drain from the tub each minute.

③ How long does it take for the tub to fully drain?

_____

Input (number of minutes)	Output (water remaining in the tub)
0	60 gallons
1	
2	
3	
4	
5	

Reflect
How many gallons of water are left in the tub 1 minute after the plug is pulled? How many gallons of water are left in the tub 3 minutes after the plug is pulled? Is this an example of same-step pattern or a changing-step pattern?

_____

_____

_____

Patterns and Relationships • Lesson 5 Review

Week 3 — Patterns and Graphs

Lesson 1

Key Idea
Patterns can be represented with pictures, rules, and tables. They can also be represented with graphs.

Try This
Below is a graph that shows how far Lisa can drive, depending on the number of gallons of gasoline in the car's tank.

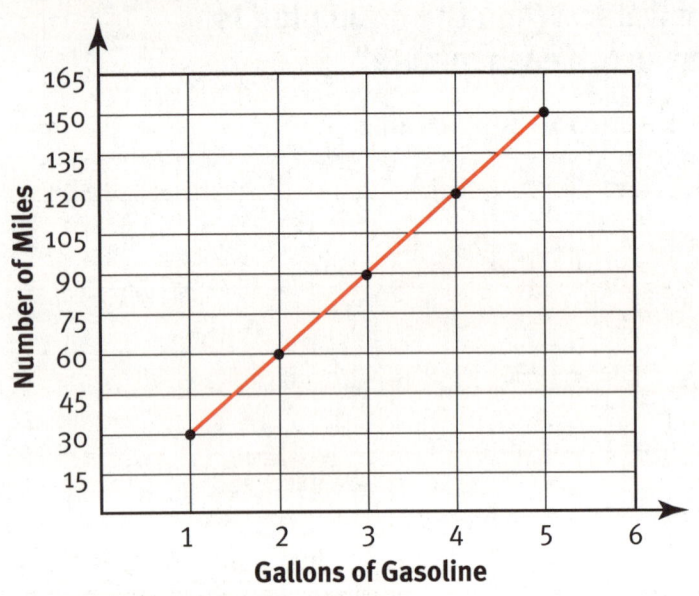

1 Which axis represents the number of miles Lisa can drive?

_____

2 Which axis represents the amount of gasoline in Lisa's car?

_____

3 As the number of gallons of gasoline increases, what happens to the distance that Lisa can drive? Is this increase a same-step increasing pattern or a changing-step pattern?

_____
_____

4 Describe the pattern shown in the graph.

_____
_____
_____
_____

22 Number Patterns and Relationships • Week 3

Practice

The graph shows the money Thomas made mowing lawns.

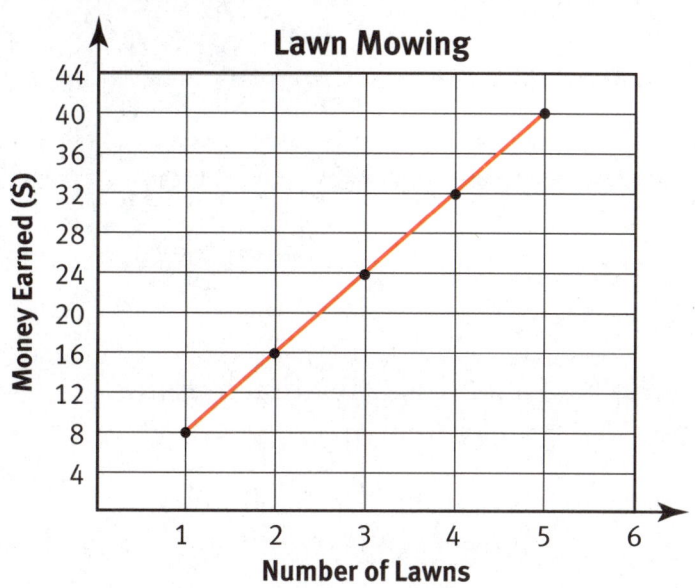

5 The data points are connected to help you see the trend of the data. As the number of lawns mowed increases, what happens to the amount of money earned?

_____

6 By connecting the data points, you are showing that the data is continuous. Should these data points be connected? Explain your answer.

_____
_____
_____
_____

Reflect
Create an input/output table, using the information above.

Week 3 — Patterns and Graphs

Lesson 2

Key Idea
When creating a graph, be sure to label the axes and give it a title.

Try This
Follow the steps to create a graph of the pattern.

Input (movie tickets)	Output (total cost)
1	$6
2	$12
3	$18
4	$24
5	$30

Step 1 Label the horizontal axis and the vertical axis.

Step 2 Plot a point for each pair of numbers in the table.

Step 3 Give your graph a title.

Practice
Mandy is babysitting for her neighbors. Graph the pattern shown in the input/output table.

Input (number of hours)	Output (money earned)
1	$5
2	$10
3	$15
4	$20
5	$25

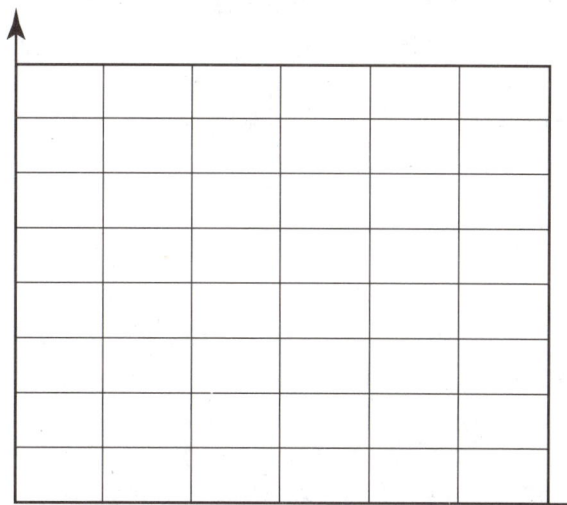

1 Describe the pattern shown in the graph.

___

2 How much does Mandy earn for babysitting 6 hours?

___

Reflect
Can you determine the rule for a pattern by just looking at the graph? Explain and give an example.

___

___

Patterns and Graphs • Lesson 2

Week 3

Patterns and Graphs

Lesson 3

> **Key Idea**
> You can use graphs to compare two related patterns.

Try This
Create a graph for each input/output table.
Answer each question.

Milk Cartons Sold

Input (day)	Output (milk sold for the week)
Monday	25 cartons
Tuesday	50 cartons
Wednesday	75 cartons
Thursday	100 cartons
Friday	125 cartons

Milk Cartons Left

Input (day)	Output (milk cartons left in the cafeteria)
Monday	125 cartons
Tuesday	100 cartons
Wednesday	75 cartons
Thursday	50 cartons
Friday	25 cartons

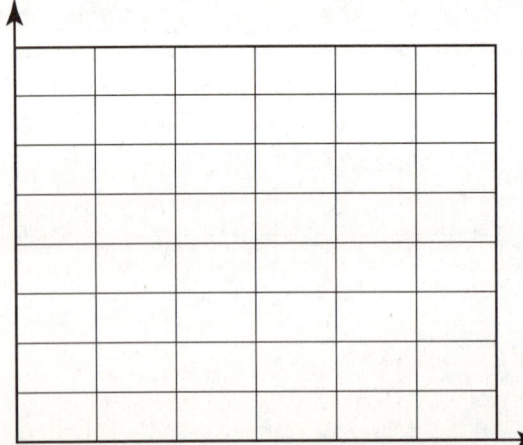
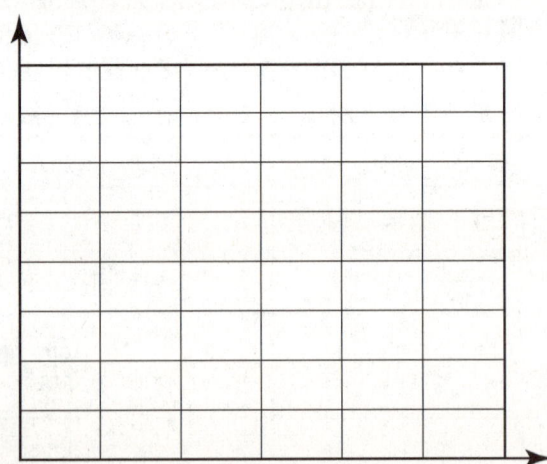

1 Which arrow line or axis represents the day of the week? Which represents the number of milk cartons left?

_____

_____

26 Number Patterns and Relationships • Week 3

Practice
Use your graphs from Try This to answer each question.

② Describe the pattern shown in the first graph.
___

③ Describe the pattern shown in the second graph.
___

④ Which of the graphs shows a growing pattern?
___

⑤ Are these graphs same-step patterns or changing-step patterns?
___

⑥ What stays the same in the first graph? What changes?
___

⑦ What stays the same in the second graph? What changes?
___

⑧ How are the two graphs related?
___
___

Reflect
Can you show both patterns on the same graph? Explain.
___
___

Patterns and Graphs • Lesson 3   **27**

Week 3

Patterns and Graphs

Lesson 4

> **Key Idea**
> You can use graphs to tell a story or make an informed decision.

Try This
Choose the story that belongs with each graph.

Story A Melissa rode her bike for 40 minutes. The table shows the distance she traveled.

Time	10 minutes	20 minutes	30 minutes	40 minutes
Distance	3 miles	6 miles	9 miles	12 miles

Story B Mrs. Swanson walked her dog for 40 minutes. The table shows the number of blocks she covered.

Time	10 minutes	20 minutes	30 minutes	40 minutes
Distance	6 blocks	12 blocks	18 blocks	24 blocks

1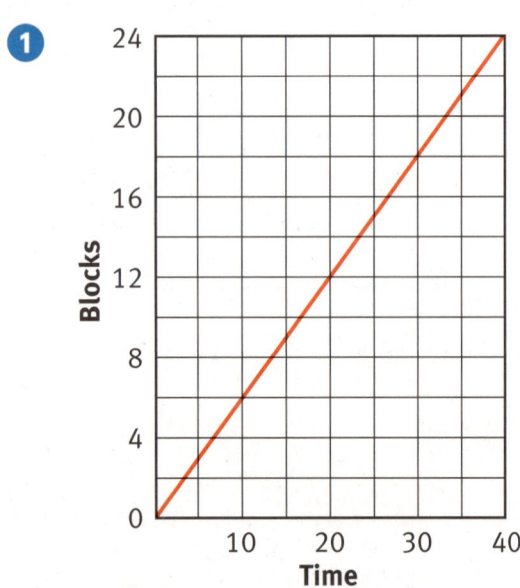

_____

2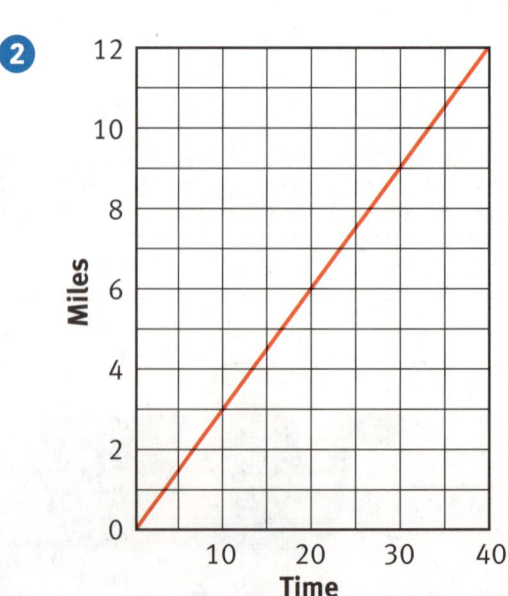

_____

28 Number Patterns and Relationships • Week 3

Practice

Create a graph for the data in the table.

3

Number of Books	1	2	4	6
New Vocabulary Words	2	4	8	12

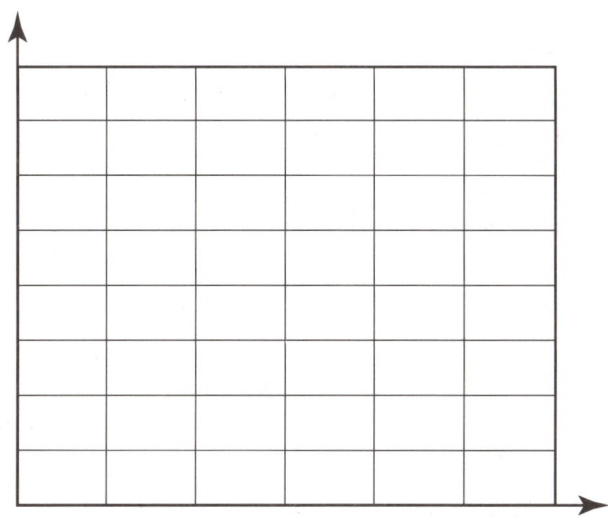

Reflect

Create a table that compares the outside temperature throughout the morning and afternoon of a winter day. Graph the data in the table.

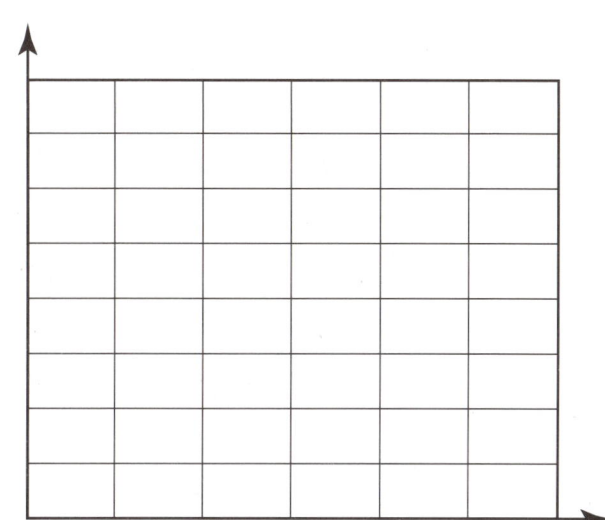

Patterns and Graphs • Lesson 4

Week 3 — Patterns and Graphs

Lesson 5 Review

This week you explored how patterns look in graphs. You used input/output tables and stories to create graphs. You also used graphs to answer questions about the pattern and data.

Lessons 1 and 2

The bookstore sells school sweatshirts for $10 each. Graph the pattern shown in the input/output table.

Input (number of sweatshirts)	Output (total cost)
1	$10
2	$20
3	$30
4	$40
5	$50

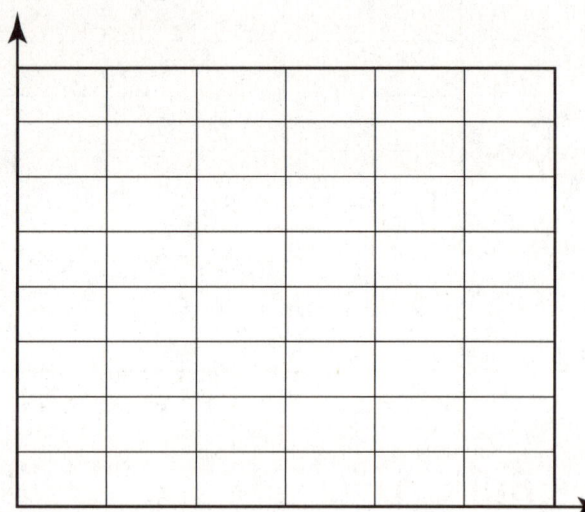

1 Describe the pattern that is shown in the graph.

_____

_____

2 How much would it cost to purchase eight sweatshirts?

_____

Reflect
What is staying the same in the graph?

_____

30 Number Patterns and Relationships • Week 3

Lesson 3 Graph the pattern shown in the input/output table.

Input (number of cars)	Output (total passengers)
1	4
2	8
3	12
4	16
5	20

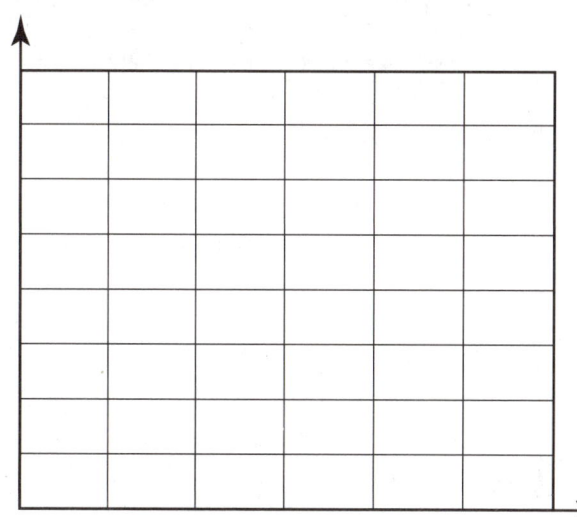

Lesson 4 Use the graph to answer each question.

❸ Which story matches the graph? Circle A or B.

A. A shoe store sold 20 pairs of shoes on Monday. The store sold no shoes on Tuesday because it was closed. On Wednesday and Thursday, 30 pairs of shoes were sold.

B. A hot air balloon rose to 250 feet. It stayed there for a while and then rose to 500 feet. After a little while longer, the balloon began its descent.

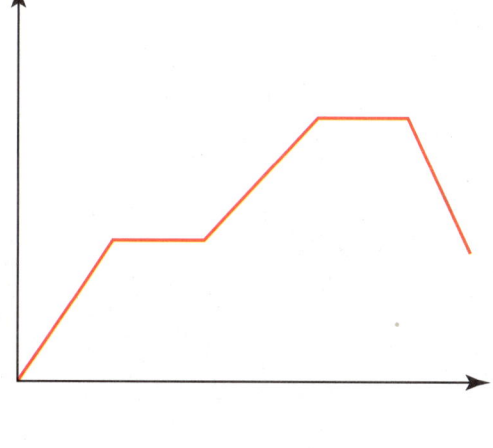

Reflect
What label would you put on the horizontal axis? What label would you put on the vertical axis?

Patterns and Graphs • Lesson 5 Review

Week 4 — Variables and Equality

Lesson 1

Key Idea

An **equation** is a number sentence which states that two mathematical expressions are equal.

$2 + 3 = 5$ $\qquad$ $11 - 4 = 7$ $\qquad$ $6 - 2 = 3 + 1$

Sometimes equations have unknown values. You can show unknown values with pictures, boxes, or letters.

$4 + \square = 12$ $\qquad$ $b - 9 = 5$

Try This

Find the unknown value in each equation. Substitute values into the equation until you have a true number sentence.

1 $\square + 6 = 8$

What is $\square$?

2 $\triangle + 1 = 7$

What is $\triangle$?

3 $4 + \triangle = 8$

What is $\triangle$?

4 $5 - \bigcirc = 2$

What is $\bigcirc$?

5 $a + 8 = 10$

$a = $ _____

6 $12 - 7 = z$

$z = $ _____

Practice

Find the unknown value in each equation. The same shapes represent the same value.

7 $\square + \square = 10$

What is $\square$?

8 $\bigcirc + \bigcirc = 2$

What is $\bigcirc$?

9 △ + △ = 6

What is △?

10 ◯ + ◯ + ◯ = 15

What is ◯?

Find the unknown value in each equation.

11 9 − ☐ = 3

What is ☐?

12 8 − n = 6

n = _____

13 t − 7 = 2

t = _____

14 14 + ◯ = 19

What is ◯?

15 x + 4 = 12

x = _____

16 17 − y = 12

y = _____

Reflect

What values of ☐ and △ make a true number sentence? Is there more than one correct answer? Explain.

☐ − △ = 3

_____
_____
_____

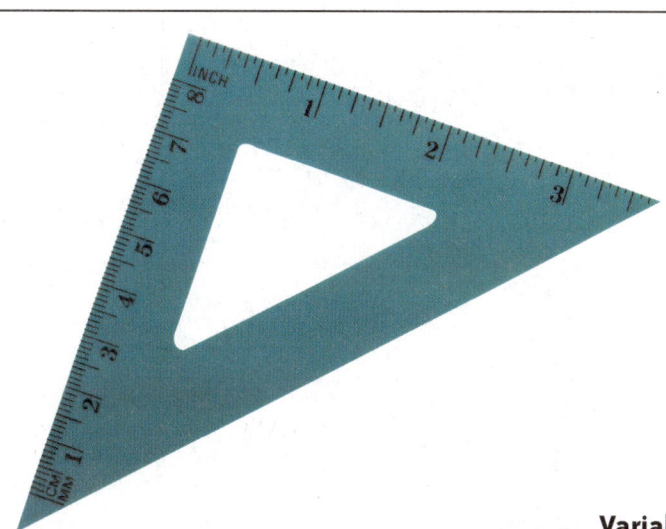

Variables and Equality • Lesson 1   33

Week 4: Variables and Equality

Lesson 2

Key Idea
You can use the idea of weights to help solve equations.

Try This
Answer each question to find the weight of the toy car.

3 pounds

8 pounds

1 How much does the piggy bank weigh?

_____

2 How much do the piggy bank and toy car weigh altogether?

_____

3 Fill in the blanks below to help you find the weight of the toy car.

Piggy bank = 3

Piggy bank + toy car = 8

_____ + toy car = 8

_____ + _____ = 8

Toy car = _____

The toy car weighs _____.

Practice
Find each unknown weight. Write a number sentence to show your work.

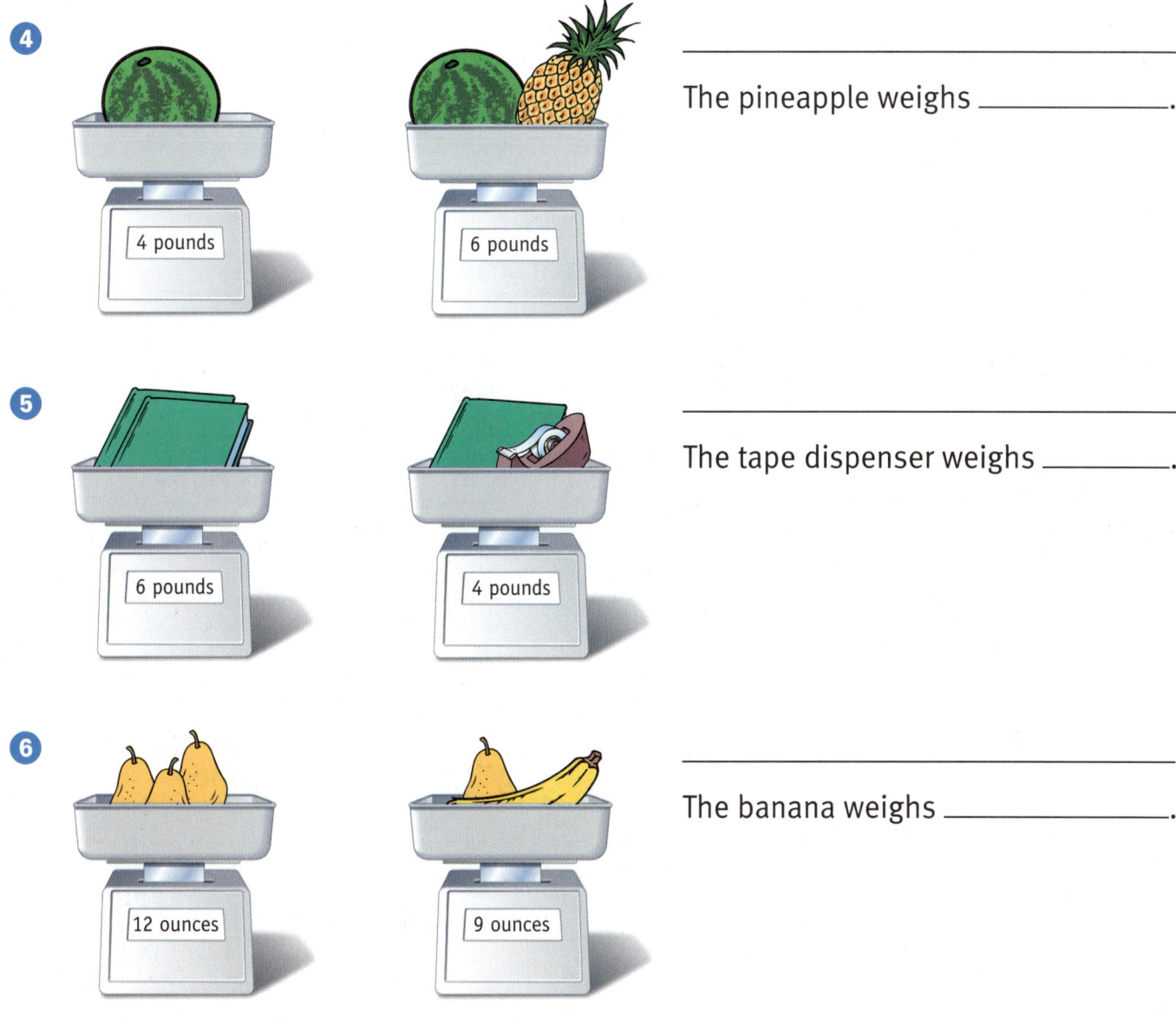

4. The pineapple weighs _____.

5. The tape dispenser weighs _____.

6. The banana weighs _____.

Reflect
How did you decide how much one pear weighs in Problem 6? Explain.

_____

_____

_____

Variables and Equality • Lesson 2

Week 4 — Variables and Equality

Lesson 3

Key Idea
Use reasoning to solve more challenging problems involving weights.

Try This
Answer each question to find the weight of each shape.

1 How much does the pyramid weigh?
_____

2 How much do the pyramid and cylinder weigh altogether?
_____

3 How much does the cylinder weigh?
_____

4 How much do the cylinder and cube weigh altogether?
_____

5 How much does the cube weigh?
_____

Practice
Find each unknown weight.

6. The shoe weighs _____.
7. The basket weighs _____.
8. The plant weighs _____.

9. Each tennis ball weighs _____.
10. Each baseball weighs _____.
11. The volleyball weighs _____.

Reflect
Would you be able to find the weight of the volleyball in Problem 7 if you had only the first and third scales? Explain your answer.

_____

Variables and Equality • Lesson 3

Week 4

Variables and Equality

Lesson 4

> **Key Idea**
> Balance scales can be used to help solve equations.
> When a scale is balanced, both sides are equal.

Try This
Use each balance scale to find two equal weights.

1.

The weight of 1 orange is the same as the weight of _____.

2.

The weight of 1 toy puppy is the same as the weight of _____.

3.

The weight of 1 box of crayons is the same as the weight of _____ _____.

4.

The weight of 4 baseballs is the same as the weight of _____ _____.

38   Number Patterns and Relationships • Week 4

Practice
Find each unknown weight. Draw your answer on the scale with the question mark.

5

6

7

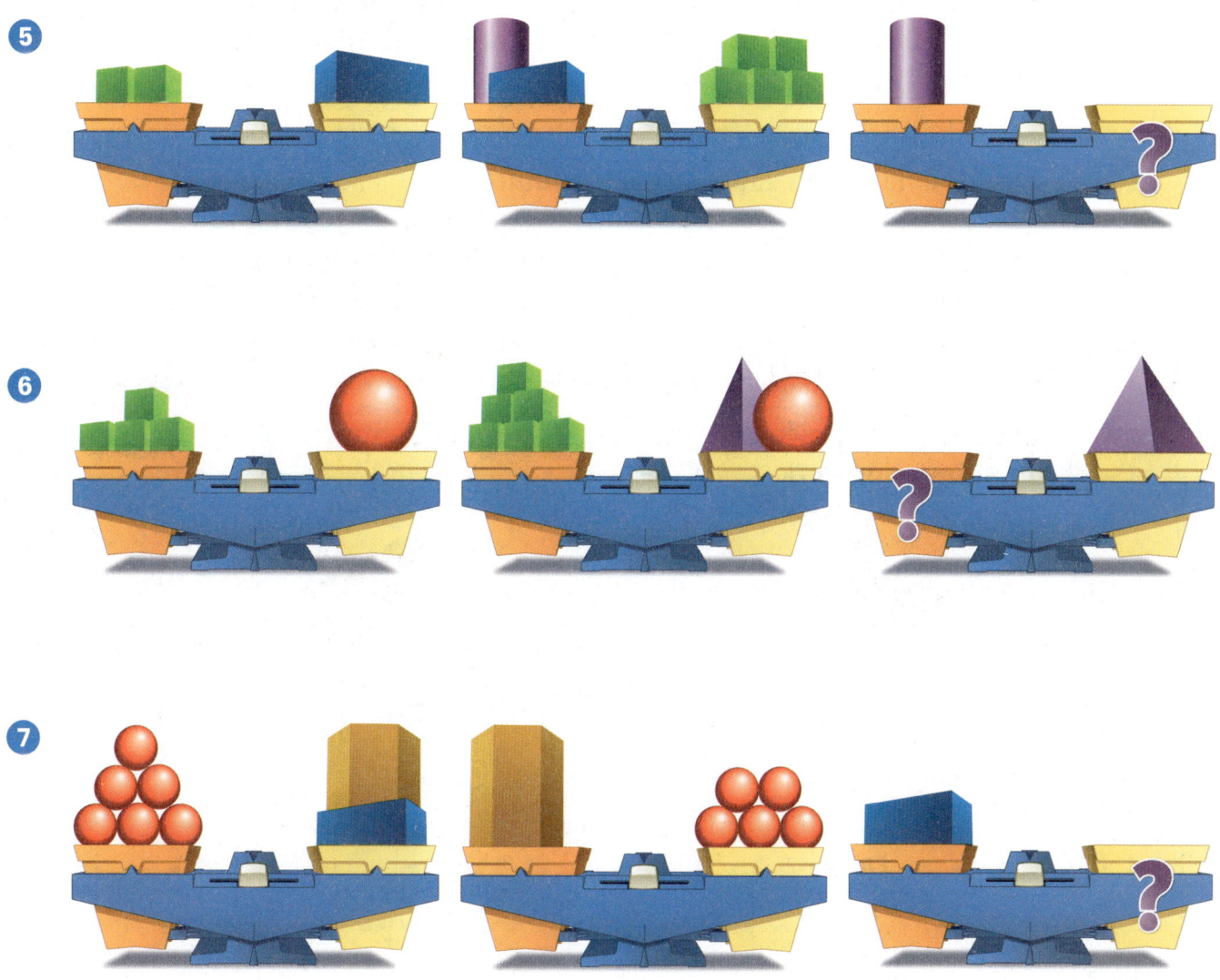

Reflect
What part of a number sentence is represented by the balance scale? Explain.

_____
_____
_____

Variables and Equality • Lesson 4   39

Week 4

Variables and Equality

Lesson 5 Review

This week you explored equality and unknown values in number sentences. You used shapes to represent missing numbers in an equation. You also related number sentences to weights and balance scales.

Lesson 1 Find the unknown value in each equation.

① $c + 2 = 8$

$c =$ _____

② $\square + \square + \square = 9$

What is $\square$?

Lesson 2 Find each unknown weight. Write a number sentence to show your work.

③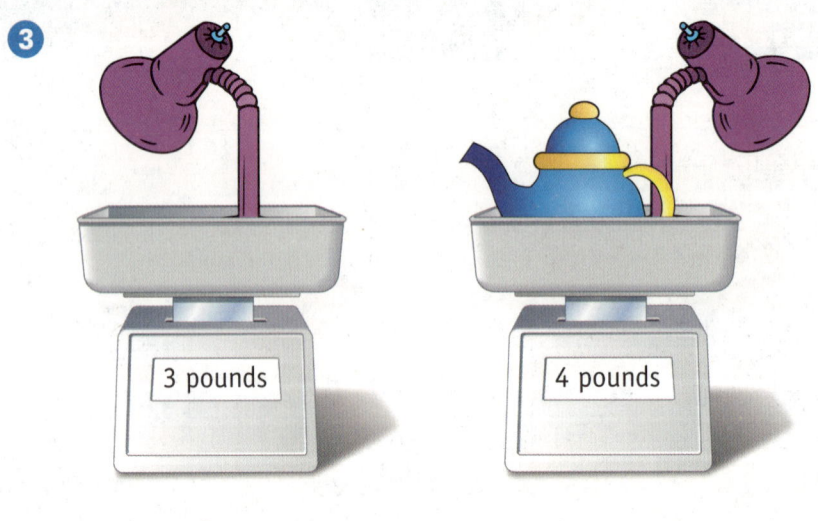

_____

The teapot weighs _____.

Reflect

Explain how to find the values of the unknowns in the number sentence.

$\square + \square + \square = 21$

_____

40 Number Patterns and Relationships • Week 4

Lesson 3

The knife weighs _____.

Lesson 4 Find the unknown weight. Draw your answer on the scale with the question mark.

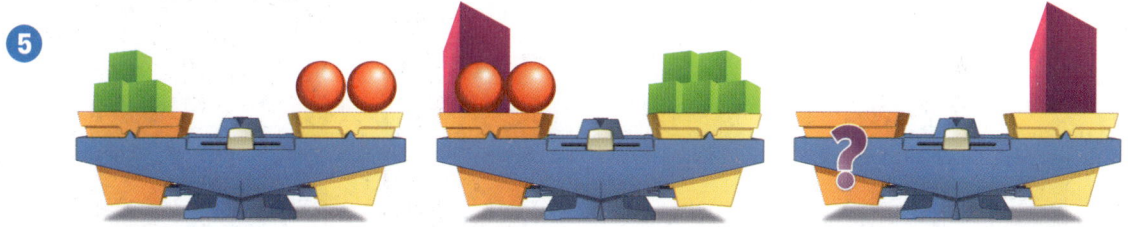

Reflect
Use shapes from Problem 5 to balance each scale. Draw your answer on the scale with the question mark.

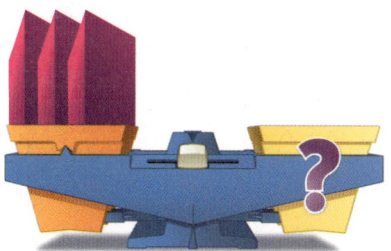 

Variables and Equality • Lesson 5 Review **41**

Week 1

Exploring Patterns

Practice

1. Draw the next set in the pattern.

Set 1   Set 2   Set 3   _____

2. Draw the missing set in the pattern.

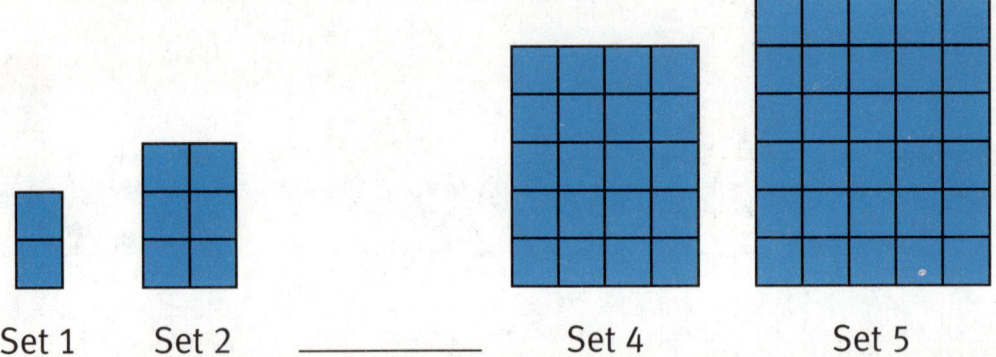

Set 1   Set 2   _____   Set 4   Set 5

3. Tell whether each pattern above is a same-step or a changing-step growing pattern.

_____

_____

4. Design a same-step or changing-step pattern. Show sets 1, 2, and 3, and describe the pattern.

42 Number Patterns and Relationships • Week 1 Practice

Week 2

Patterns and Relationships

Practice

Complete the table for the pattern shown below.

1

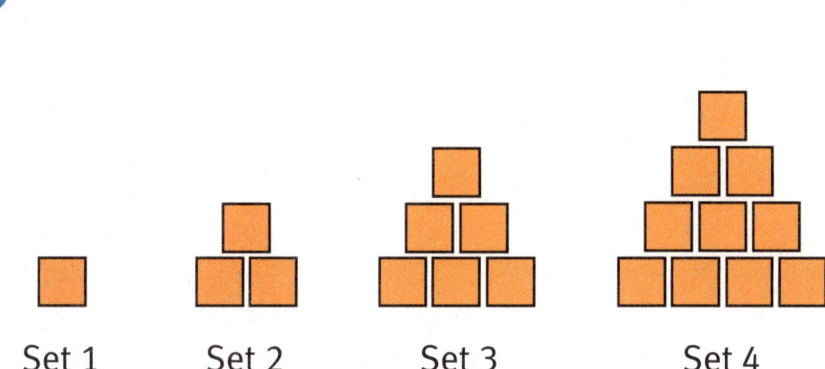

Set 1   Set 2   Set 3   Set 4

Set	Number of Squares

Complete each input/output table.

2 Candy bars cost 50¢ each.

Input (number of candy bars)	Output (total cost)

3 Jean can ride her bike 20 miles per hour.

Input (number of hours)	Output (number of miles)

Number Patterns and Relationships • Week 2 Practice 43

Week 3 Patterns and Graphs

Practice

1 The Booster Club sells gourmet cookies for $1.50 each. Complete the input/output table, and graph the pattern.

Input (number of cookies)	Output (total cost)

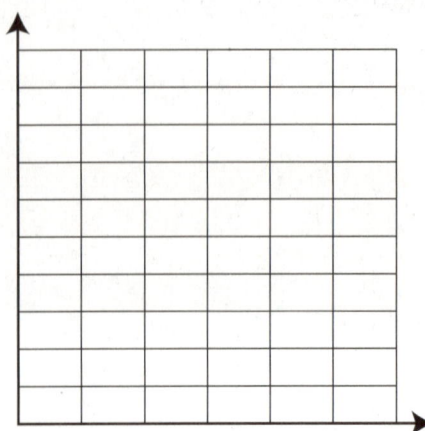

2 Describe the pattern that is shown in the graph.

_____
_____
_____

3 How much would it cost to purchase 10 cookies?

_____

4 What remains the same in the graph?

_____

44 Number Patterns and Relationships • Week 3 Practice

Unit 2 Workbook
Level **F**

NUMBER WORLDS

Number Patterns and Relationships

featuring
building Blocks
Software

Author
Sharon Griffin
*Associate Professor of Education and
Adjunct Associate Professor of Psychology*
Clark University
Worcester, Massachusetts

Building Blocks Authors

Douglas H. Clements
*Professor of Early Childhood
and Mathematics Education*
University at Buffalo
State University of New York, New York

Julie Sarama
Associate Professor of Mathematics Education
University at Buffalo
State University of New York, New York

Contributing Writers
Sherry Booth, *Math Curriculum Developer,* Raleigh, North Carolina
Elizabeth Jimenez, *English Language Learner Consultant,* Pomona, California

Program Reviewers

Jean Delwiche
Almaden Country School
San Jose, California

Cheryl Glorioso
Santa Ana Unified School District
Santa Ana, California

Sharon LaPoint
School District of Indian River County
Vero Beach, Florida

Leigh Lidrbauch
Pasadena Independent School District
Pasadena, Texas

Dave Maresh
Morongo Unified School District
Yucca Valley, California

Mary Mayberry
Mon Valley Education Consortium, AIU 3
Clairton, Pennsylvania

Lauren Parente
Mountain Lakes School District
Mountain Lakes, New Jersey

Juan Regalado
Houston Independent School District
Houston, Texas

M. Kate Thiry
Dublin City School District
Dublin, Ohio

Susan C. Vohrer
Baltimore County Public Schools
Baltimore, Maryland

SRAonline.com

Copyright © 2007 SRA/McGraw-Hill.

All rights reserved. Except as permitted under the United States Copyright Act, no part of this publication may be reproduced or distributed in any form or by any means, or stored in a database or retrieval system, without the prior written permission of the publisher, unless otherwise indicated.

Printed in the United States of America.

Send all inquiries to:
SRA/McGraw-Hill
4400 Easton Commons
Columbus, OH 43219

R53180.01

7 8 9 QPE 12 11 10 09

Photo Credits
2–14 ©PhotoDisc/Getty Images, Inc.;
15 ©Stockbyte/Stockbyte; **16–23** ©PhotoDisc/Getty Images, Inc.; **24** ©Eyewire/Getty Images, Inc.; **25** ©Matt Meadows; **33** ©Stockbyte/Stockbyte

Contents

Number Patterns and Relationships

Week 1 Exploring Patterns 2

Week 2 Patterns and Relationships 12

Week 3 Patterns and Graphs 22

Week 4 Variables and Equality 32

Week 1 Practice 42

Week 2 Practice 43

Week 3 Practice 44

Week 4 Practice 45

Week 1 — **Exploring Patterns**

Lesson 1

Key Idea

Same-step patterns are patterns in which the amount of change or growth is the same from one set to the next.

Ask the following questions to help you look for a pattern.

- What is changing from one set to the next?
- What stays the same from one set to the next?
- How would you create the next set?

Try This

Describe what changes in each pattern. Sketch and label the next figure in the pattern.

1

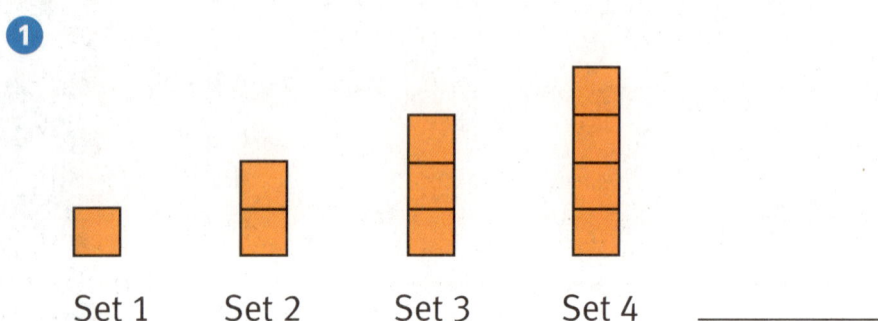

Set 1   Set 2   Set 3   Set 4   _____

_____

_____

2

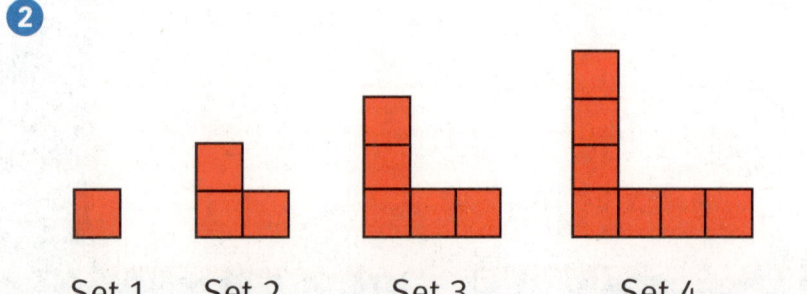

Set 1   Set 2   Set 3   Set 4   _____

_____

_____

Practice
Sketch and label the next two terms in each pattern.

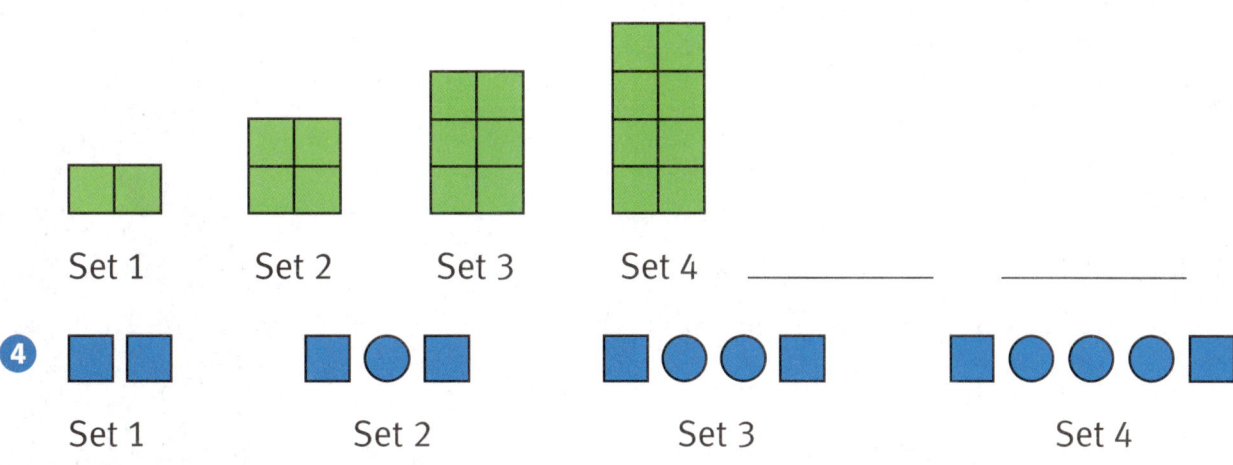

Reflect
Mandy had 5 pennies in her piggy bank on Monday. Each day she put 10 more pennies into the bank. How much money will she have in the piggy bank on Friday? Explain.

Monday	Tuesday	Wednesday	Thursday	Friday
5¢	15¢	25¢	35¢	_____

_____
_____
_____
_____

Exploring Patterns • Lesson 1

Week 1 — Exploring Patterns

Lesson 2

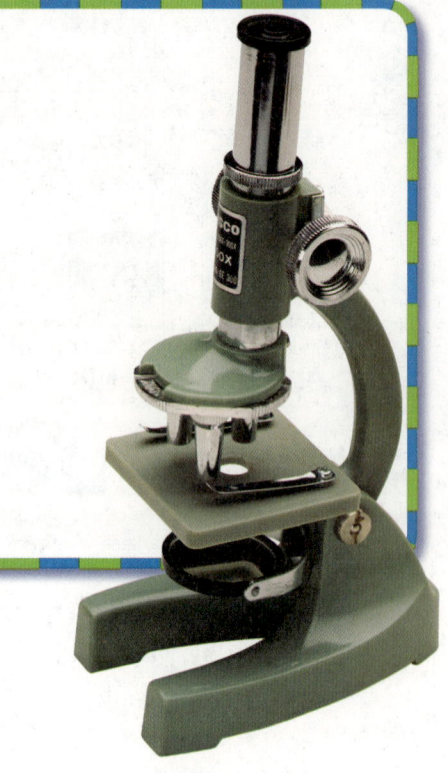

Key Idea

Changing-step patterns are patterns in which the amount of change or growth changes in a regular and predictable way from one set to the next.

Ask the following questions to help you identify a changing-step pattern:

- What changes from one set to the next?
- What stays the same from one set to the next?
- How would you create the next set?

Try This

Describe the changes in each pattern. Then sketch and label the next set in the pattern.

1

  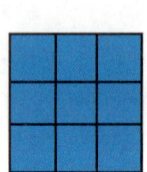 

Set 1   Set 2   Set 3   Set 4   _____

_____

_____

2

Set 1   Set 2   Set 3   _____

_____

_____

_____

Practice

Sketch and label the next term in the pattern.

3

Set 1   Set 2   Set 3   Set 4   _____

4

Set 1   Set 2   Set 3   Set 4   _____

Reflect

Look for a pattern in the table. If the pattern continues, how many chin-ups will Jarrod do on Friday? Explain your answer.

Jarrod's Chin-up Chart				
Monday	Tuesday	Wednesday	Thursday	Friday
8 chin-ups	10 chin-ups	13 chin-ups	17 chin-ups	

_____

_____

_____

Week 1 — Exploring Patterns

Lesson 3

Key Idea
You can use clues from the sets given in a pattern to identify missing sets.

Try This
Look for the changes in each pattern. Then draw and label the missing set.

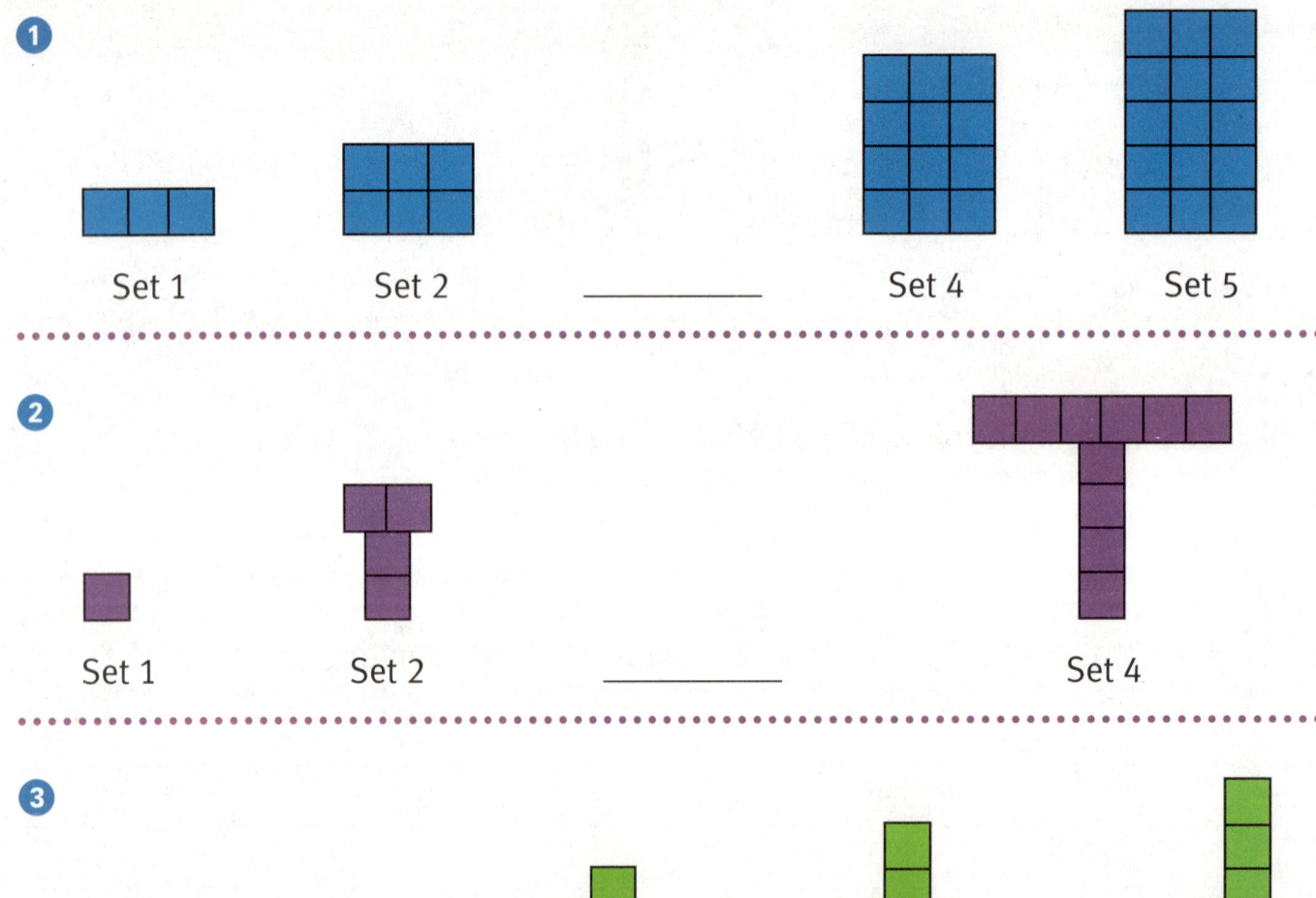

Practice

Look for the changes in each pattern. Tell whether it is a same-step growing pattern or a changing-step growing pattern. Then draw and label the missing set.

4.

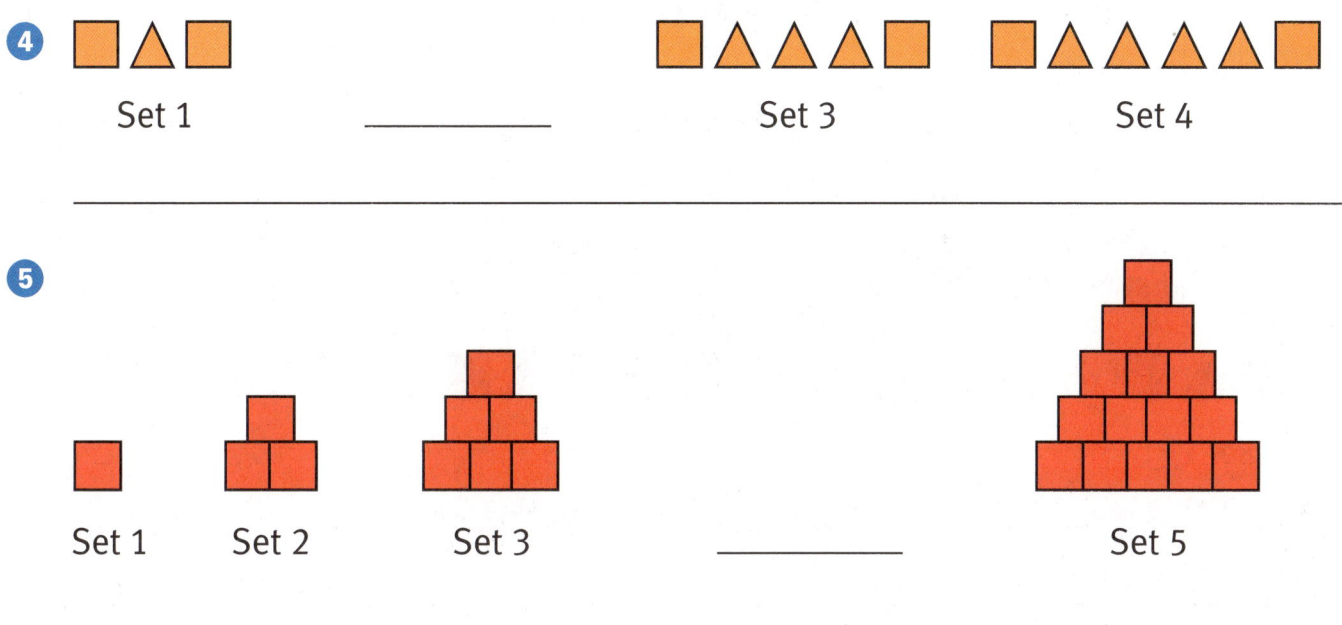

Set 1   _____   Set 3   Set 4

5.

Set 1   Set 2   Set 3   _____   Set 5

Reflect

Do you think it is easier to find a missing set in a same-step or a changing-step growing pattern? Explain.

_____
_____
_____
_____

Exploring Patterns • Lesson 3

Week 1 Exploring Patterns

Lesson 4

Key Idea
You can use pattern blocks to create your own growing patterns.

Try This
Draw and label the next set for each pattern. Use the pattern to answer the questions.

1.

   Set 1   Set 2   Set 3   _____

   a. Is this a same-step or a changing-step growing pattern?

   _____

   b. How is the pattern changing from set to set?

   _____

2.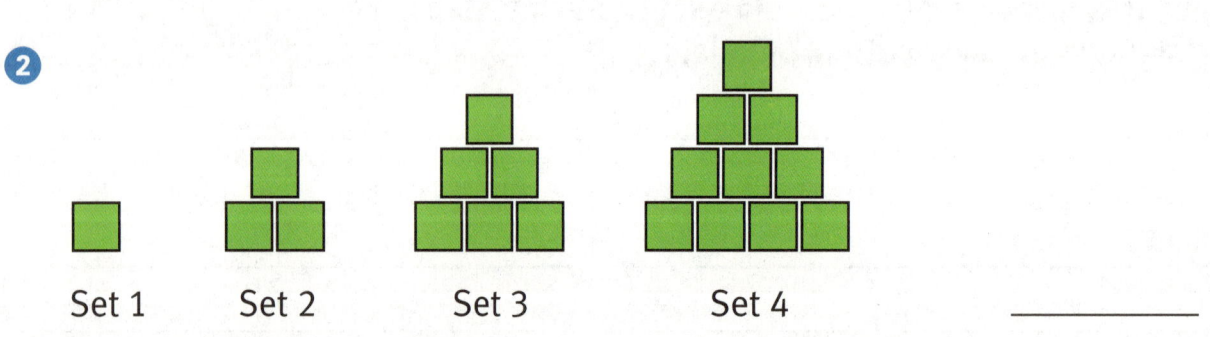

   Set 1   Set 2   Set 3   Set 4   _____

   a. Is this a same-step or a changing-step growing pattern?

   _____

   b. How is the pattern changing from set to set?

   _____

   _____

8 Number Patterns and Relationships • Week 1

Practice
Draw your own pattern in the space below. Exchange your pattern with a partner, and have your partner answer each question.

❸ Is this a same-step or a changing-step growing pattern?
_____

❹ How is the pattern changing from set to set?
_____

❺ What would the next set of the pattern look like? Describe it.
_____

Reflect
Use different shapes to create the first few sets of a pattern. Explain how the pattern changes from set to set.

_____
_____

Exploring Patterns • Lesson 4

Week 1 — Exploring Patterns

Lesson 5 Review

This week you explored patterns. You examined same-step patterns and changing-step patterns.

Lesson 1   Sketch and label the next set in the pattern.

1

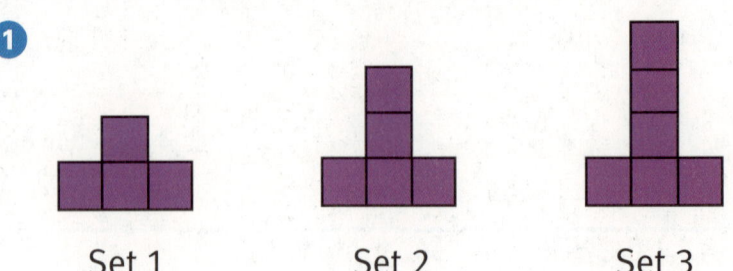

Set 1   Set 2   Set 3   _____

2

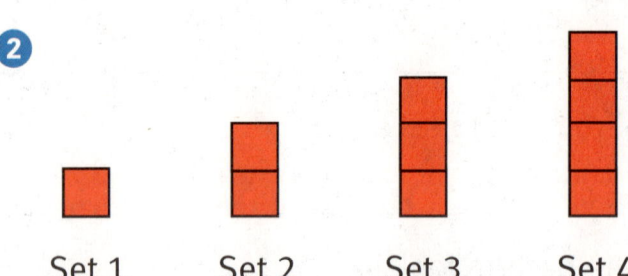

Set 1   Set 2   Set 3   Set 4   _____

Lesson 2   Sketch and label the next set in the pattern.

3

Set 1   Set 2   Set 3   _____

Reflect

Tell whether each pattern above is a same-step or a changing-step growing pattern.

_____

_____

_____

Lesson 3 Sketch and label the missing set in the pattern.

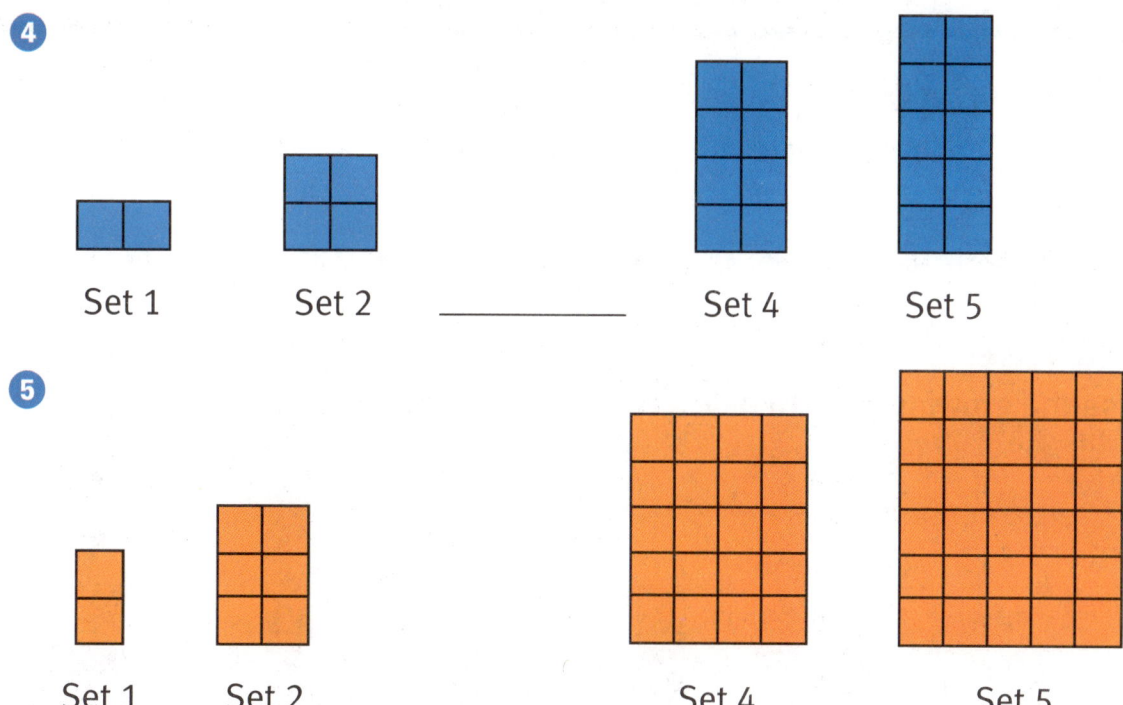

Lesson 4 ⑥ Use pattern blocks to design Set 1 of a pattern. Then show Sets 2 and 3.

Reflect
Describe how to find the missing set. Then draw and label the missing set.

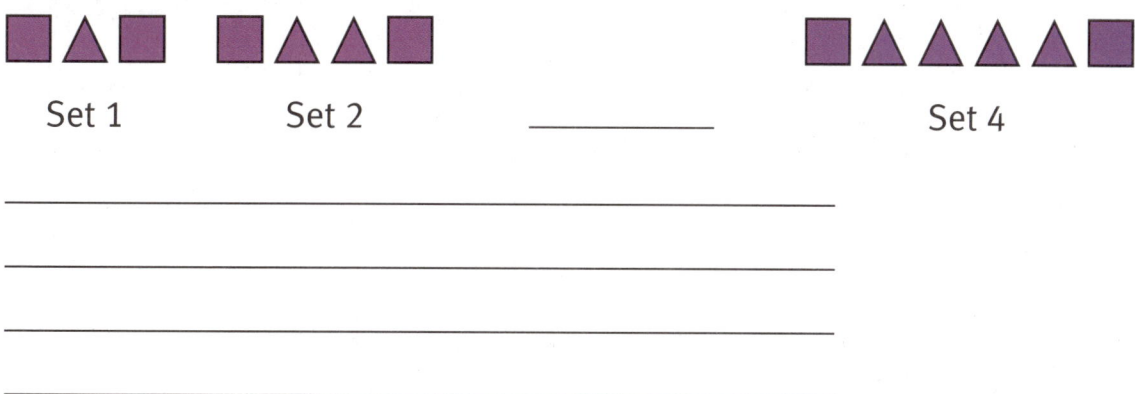

_____

_____

_____

_____

Exploring Patterns • Lesson 5 Review 11

Week 2 Lesson 1

Patterns and Relationships

Key Idea
You can use numbers to represent the growth in patterns.

Try This
Use the growing pattern below to answer each question.

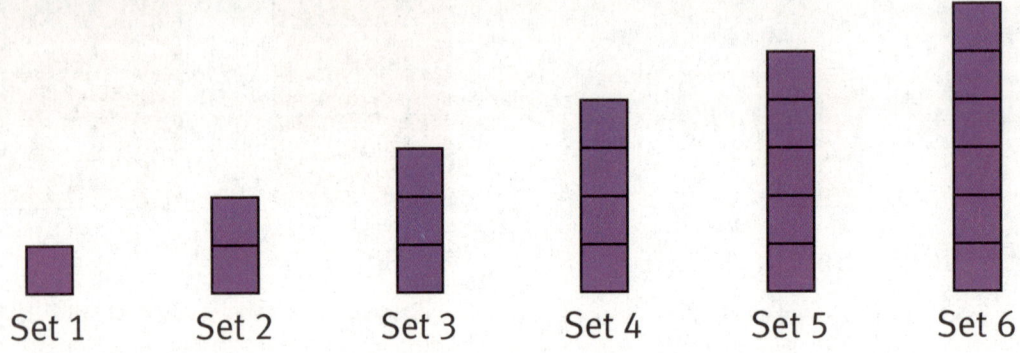

Set 1   Set 2   Set 3   Set 4   Set 5   Set 6

1 Use the pattern to complete the table below.

Set	Number of Squares
1	
2	
3	
4	
5	
6	

2 What patterns do you notice in the table between the set number and the number of squares?

_____

_____

3 Use the pattern from Problem 2 to predict the number of squares in the tenth set without building or drawing all the sets up to the tenth.

_____

Practice
Use the pattern to complete the table.

4

Set 1   Set 2   Set 3    Set 4   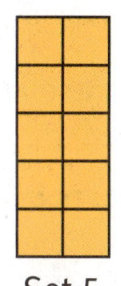 Set 5

Set	Number of Squares
1	
2	
3	
4	
5	

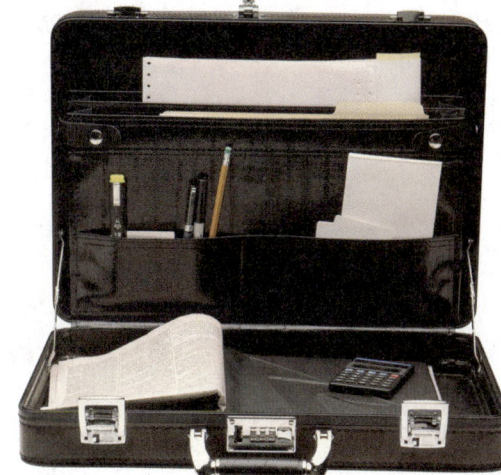

5 What patterns do you notice in the table above between the set number and the number of squares?

_____

6 Use the pattern from Problem 5 to predict the number of squares in the tenth set without building or drawing all of the sets up to the tenth.

_____

Reflect
When you use a pattern table to help you create a pattern with shapes, what is the smallest number of sets you need to figure out the pattern? Explain.

_____
_____
_____

Patterns and Relationships • Lesson 1

Week 2 — Patterns and Relationships

Lesson 2

Key Idea
You can use numbers to represent the growth in patterns.

Try This
Use the growing pattern below to answer each question.

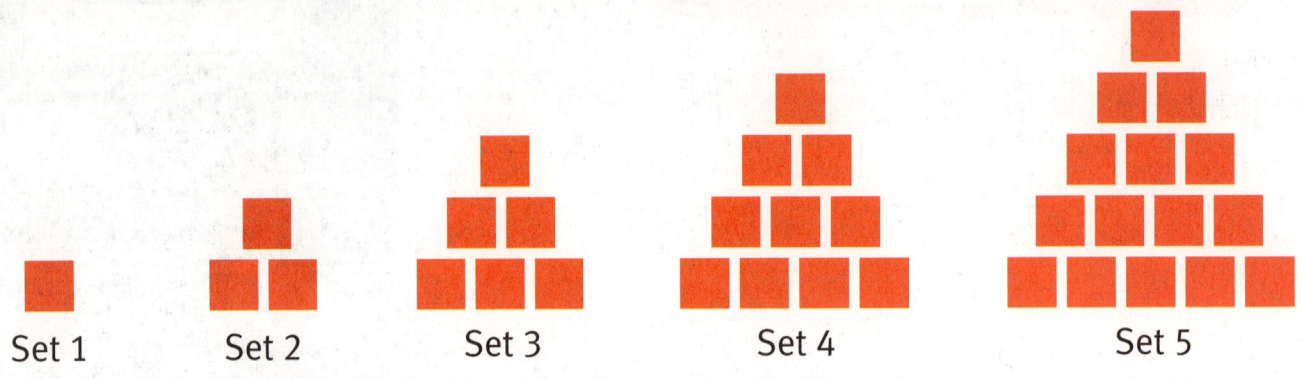

Set 1   Set 2   Set 3   Set 4   Set 5

1. Is this a same-step or a changing-step growth pattern?
   _____

2. Describe the growth shown in the pattern.
   _____

3. How many squares were used to create each set?
   _____

4. Use your answers to complete the table below.

Set	Number of Squares
1	
2	
3	
4	
5	

Practice
Use the pattern to complete the table.

5

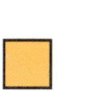    

Set 1   Set 2   Set 3   Set 4   Set 5

Set	Number of Squares
1	
2	
3	
4	
5	

6 What patterns do you notice in the table above between the set number and the number of squares?

_____

7 Use the pattern to predict the number of squares in the tenth set without building or drawing all of the sets up to the tenth.

_____

Reflect
What do you notice about the table for a changing-step pattern? How does it compare with the table for a same-step pattern?

_____

_____

_____

Patterns and Relationships • Lesson 2

Week 2 · Lesson 3
Patterns and Relationships

> **Key Idea**
>
> The pattern tables are examples of input/output tables.
>
> For each input value (the set number), there is a certain output value (number of squares).
>
> Input/output tables can also be used to model real-world situations.

Try This

Thomas mows lawns in his neighborhood to earn money. He earns $8 for each lawn he mows. Use the input/output table to answer each question.

Input (lawns mowed)	Output (money earned)
1	$8
2	$16
3	$24
4	$32
5	$40

1 How much money would Thomas earn if he mowed 6 lawns? Explain how you found your answer.

_____

_____

_____

2 Write a mathematical rule for determining the amount of money Thomas will earn for any given number of lawns mowed.

_____

Practice
Complete the table and answer each question.

3 Movie tickets cost $6 each. Complete the input/output table.

Input (number of tickets)	Output (total cost)
1	$6
2	
3	
4	
5	

4 How much would it cost to buy 4 movie tickets?

_____

5 Write a mathematical rule for determining the total cost of tickets for any given number of tickets.

_____

_____

6 For every hour Lisa drives, she uses 2 gallons of gasoline. Her gas tank holds 18 gallons when it is full. Complete the input/output table.

Input (hours of driving)	Output (gas remaining in her tank)
1	16 gallons
2	14 gallons
3	
4	
5	

7 How much gasoline is in Lisa's tank after 5 hours of driving?

_____

8 Write a mathematical rule for determining the amount of gas remaining for any given number of hours driven.

_____

_____

Reflect
What is different about the lawn mowing input/output table and the gasoline input/output table?

_____

_____

Patterns and Relationships • Lesson 3 **17**

Week 2 — Patterns and Relationships

Lesson 4

> **Key Idea**
> You can use input/output tables to help you make choices.

Try This

Anna's neighbors have hired her to pet sit their dog for seven days. They have offered two different options for being paid.

- **Option 1:** Anna receives $10 for the first day and an additional $2 per day after the first day.

- **Option 2:** Anna receives $1 for the first day. Every day after the first day she receives an additional amount that is $1 more than the previous day.

1 Complete the input/output tables for each option.

Option 1	
Day	Total Amount Earned
1	$10
2	$12
3	
4	
5	
6	
7	

Option 2	
Day	Total Amount Earned
1	$1
2	$3
3	$6
4	
5	
6	
7	

2 If Anna chooses Option 1, how much money will she be paid? If Anna chooses Option 2, how much money will she be paid?

Practice

Jim was hired to do yard work for his neighbor. The neighbor expects the work to last 5 days, but it could last 7 days. Payment options are as follows:

- **Option 1:** Jim receives $12 for the first day and $2 per day after the first day.

- **Option 2:** Jim receives $2 for the first two days. Every day thereafter he receives an amount that is $1 more than the previous day.

Complete the tables to find Jim's total earnings for the week for each option.

③ Complete the table for each option.

Option 1				Option 2		
Day	Amount Earned for the Day	Total Earnings		Day	Amount Earned for the Day	Total Earnings
1	$12			1	$2	
2				2		
3				3		
4				4		
5				5		
6				6		
7				7		

④ How much will Jim earn for 7 days if he chooses Option 1? _____

⑤ How much will Jim earn for 7 days if he chooses Option 2? _____

⑥ If the job is for only 5 days, which option will pay better? _____

Reflect

What kind of growth pattern is shown by Option 1? What kind of growth pattern is represented by Option 2?

Patterns and Relationships • Lesson 4

Week 2 — Patterns and Relationships

Lesson 5 Review

This week you explored patterns and relationships. You looked at how visual patterns can be related to number patterns. You also learned about input/output tables and solving problems.

Lessons 1 and 2

Complete the table for the pattern shown below.

1

Set 1   Set 2   Set 3   Set 4

Set	Number of Cubes
1	
2	
3	
4	

Reflect
How is the pattern changing?

_____

Lesson 3 Complete the input/output table.

❷ The bookstore sells pencils for 15¢ each.

Input (number of pencils)	Output (total cost)
1	15¢
2	
3	
4	

Lesson 4 A bathtub holds 60 gallons of water. When the drain plug is pulled, 12 gallons drain from the tub each minute.

❸ How long does it take for the tub to fully drain?

_____

Input (number of minutes)	Output (water remaining in the tub)
0	60 gallons
1	
2	
3	
4	
5	

Reflect
How many gallons of water are left in the tub 1 minute after the plug is pulled? How many gallons of water are left in the tub 3 minutes after the plug is pulled? Is this an example of same-step pattern or a changing-step pattern?

_____

_____

_____

Week 3

Patterns and Graphs

Lesson 1

> **Key Idea**
> Patterns can be represented with pictures, rules, and tables. They can also be represented with graphs.

Try This

Below is a graph that shows how far Lisa can drive, depending on the number of gallons of gasoline in the car's tank.

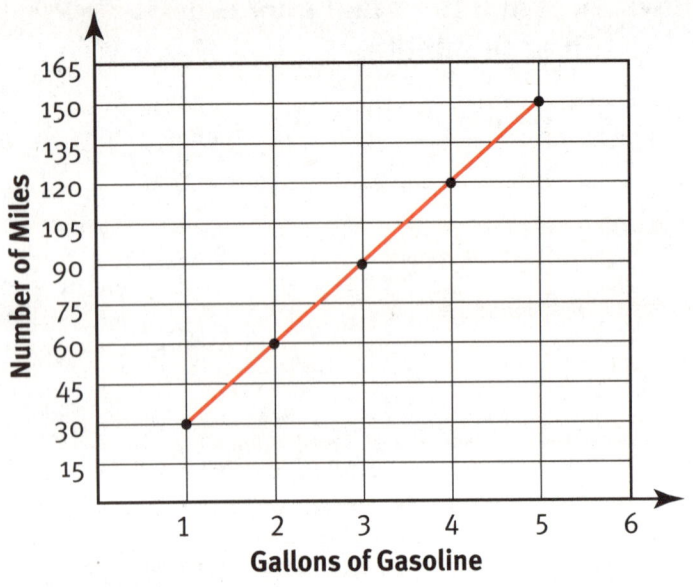

1. Which axis represents the number of miles Lisa can drive?

   _____

2. Which axis represents the amount of gasoline in Lisa's car?

   _____

3. As the number of gallons of gasoline increases, what happens to the distance that Lisa can drive? Is this increase a same-step increasing pattern or a changing-step pattern?

   _____
   _____

4. Describe the pattern shown in the graph.

   _____
   _____
   _____

22 Number Patterns and Relationships • Week 3

Practice
The graph shows the money Thomas made mowing lawns.

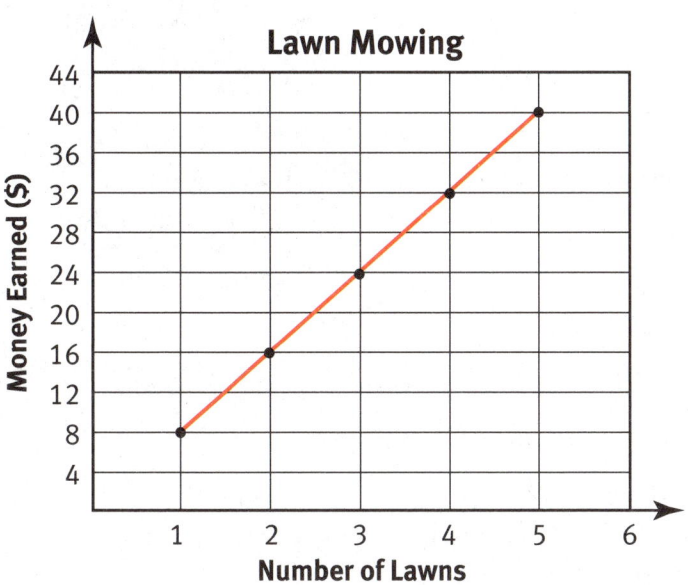

5 The data points are connected to help you see the trend of the data. As the number of lawns mowed increases, what happens to the amount of money earned?

_____

6 By connecting the data points, you are showing that the data is continuous. Should these data points be connected? Explain your answer.

_____
_____
_____
_____

Reflect
Create an input/output table, using the information above.

Week 3 — Patterns and Graphs

Lesson 2

Key Idea
When creating a graph, be sure to label the axes and give it a title.

Try This
Follow the steps to create a graph of the pattern.

Input (movie tickets)	Output (total cost)
1	$6
2	$12
3	$18
4	$24
5	$30

Step 1 Label the horizontal axis and the vertical axis.

Step 2 Plot a point for each pair of numbers in the table.

Step 3 Give your graph a title.

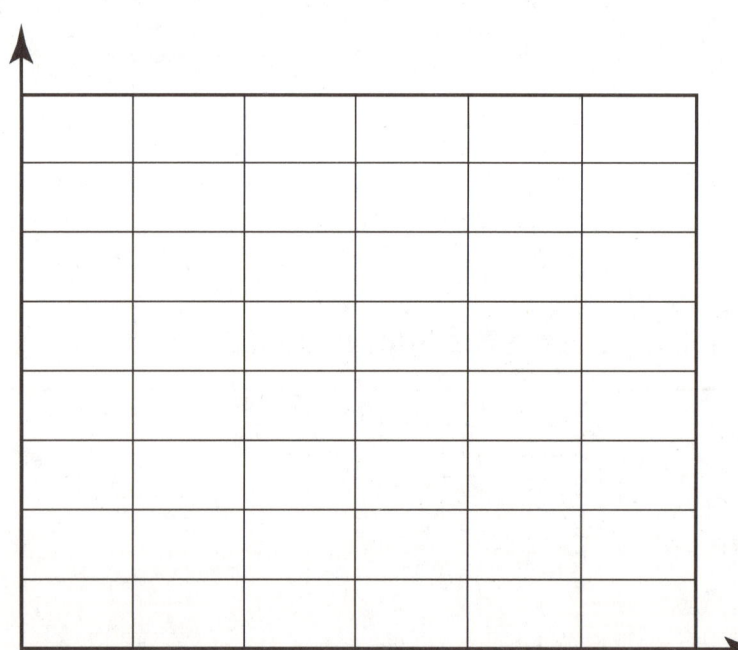

Practice

Mandy is babysitting for her neighbors. Graph the pattern shown in the input/output table.

Input (number of hours)	Output (money earned)
1	$5
2	$10
3	$15
4	$20
5	$25

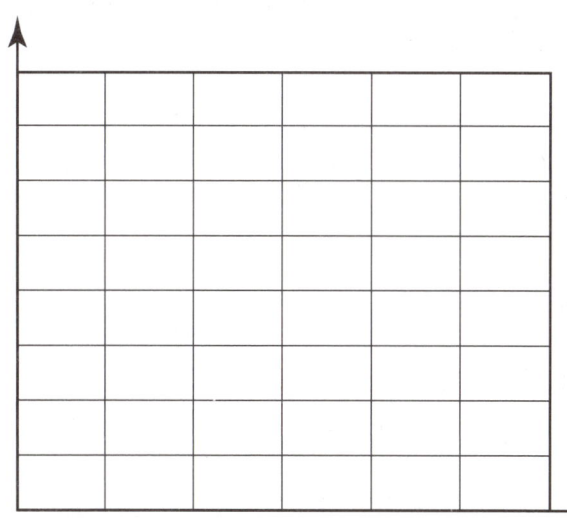

① Describe the pattern shown in the graph.

_____

② How much does Mandy earn for babysitting 6 hours?

_____

Reflect

Can you determine the rule for a pattern by just looking at the graph? Explain and give an example.

_____
_____
_____

Patterns and Graphs • Lesson 2

Week 3

Patterns and Graphs

Lesson 3

> **Key Idea**
> You can use graphs to compare two related patterns.

Try This
Create a graph for each input/output table.
Answer each question.

Milk Cartons Sold	
Input (day)	Output (milk sold for the week)
Monday	25 cartons
Tuesday	50 cartons
Wednesday	75 cartons
Thursday	100 cartons
Friday	125 cartons

Milk Cartons Left	
Input (day)	Output (milk cartons left in the cafeteria)
Monday	125 cartons
Tuesday	100 cartons
Wednesday	75 cartons
Thursday	50 cartons
Friday	25 cartons

1 Which arrow line or axis represents the day of the week? Which represents the number of milk cartons left?

___

26 Number Patterns and Relationships • Week 3

Practice
Use your graphs from Try This to answer each question.

② Describe the pattern shown in the first graph.
_____

③ Describe the pattern shown in the second graph.
_____

④ Which of the graphs shows a growing pattern?
_____

⑤ Are these graphs same-step patterns or changing-step patterns?
_____

⑥ What stays the same in the first graph? What changes?
_____

⑦ What stays the same in the second graph? What changes?
_____

⑧ How are the two graphs related?
_____
_____

Reflect
Can you show both patterns on the same graph? Explain.
_____
_____

Patterns and Graphs • Lesson 3

Week 3 Patterns and Graphs

Lesson 4

> **Key Idea**
> You can use graphs to tell a story or make an informed decision.

Try This
Choose the story that belongs with each graph.

Story A Melissa rode her bike for 40 minutes. The table shows the distance she traveled.

Time	10 minutes	20 minutes	30 minutes	40 minutes
Distance	3 miles	6 miles	9 miles	12 miles

Story B Mrs. Swanson walked her dog for 40 minutes. The table shows the number of blocks she covered.

Time	10 minutes	20 minutes	30 minutes	40 minutes
Distance	6 blocks	12 blocks	18 blocks	24 blocks

1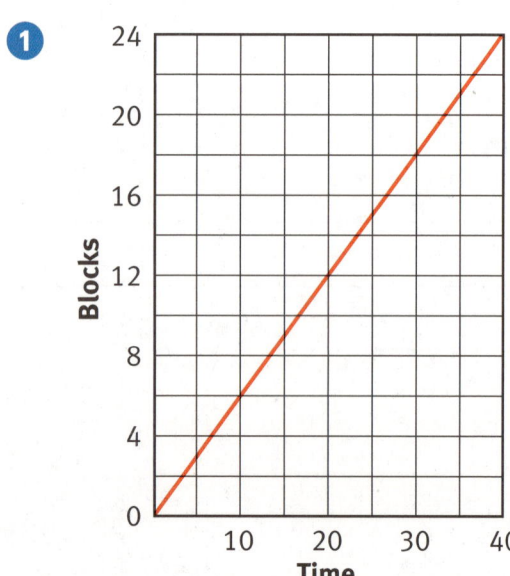

_____

2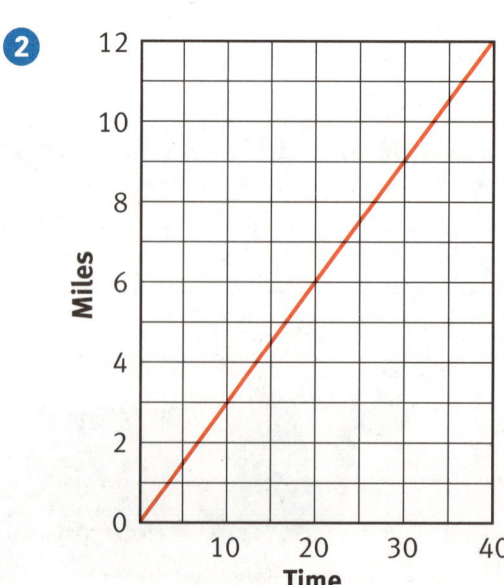

_____

28 Number Patterns and Relationships • Week 3

Practice
Create a graph for the data in the table.

3.

Number of Books	1	2	4	6
New Vocabulary Words	2	4	8	12

Reflect
Create a table that compares the outside temperature throughout the morning and afternoon of a winter day. Graph the data in the table.

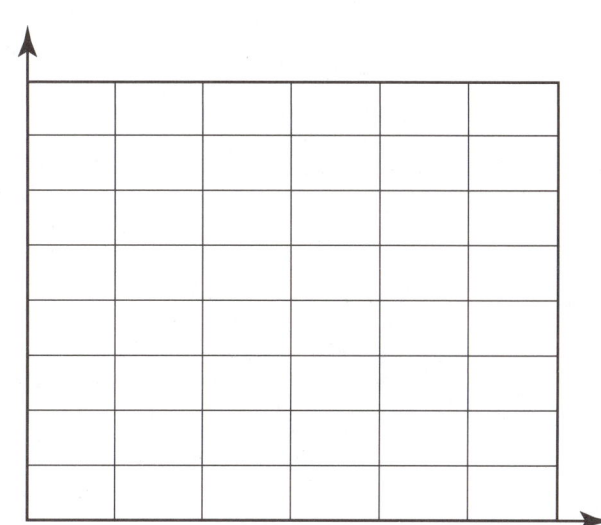

Patterns and Graphs • Lesson 4   29

Patterns and Graphs

Lesson 5 Review

This week you explored how patterns look in graphs. You used input/output tables and stories to create graphs. You also used graphs to answer questions about the pattern and data.

Lessons 1 and 2

The bookstore sells school sweatshirts for $10 each. Graph the pattern shown in the input/output table.

Input (number of sweatshirts)	Output (total cost)
1	$10
2	$20
3	$30
4	$40
5	$50

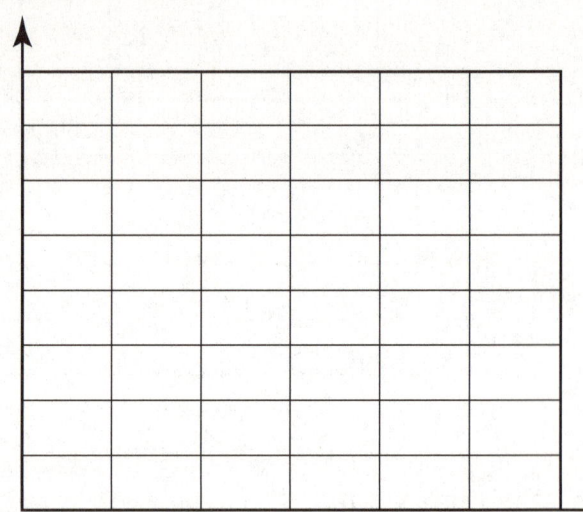

① Describe the pattern that is shown in the graph.

_____
_____

② How much would it cost to purchase eight sweatshirts?

_____

Reflect
What is staying the same in the graph?

_____

30 Number Patterns and Relationships • Week 3

Lesson 3 Graph the pattern shown in the input/output table.

Input (number of cars)	Output (total passengers)
1	4
2	8
3	12
4	16
5	20

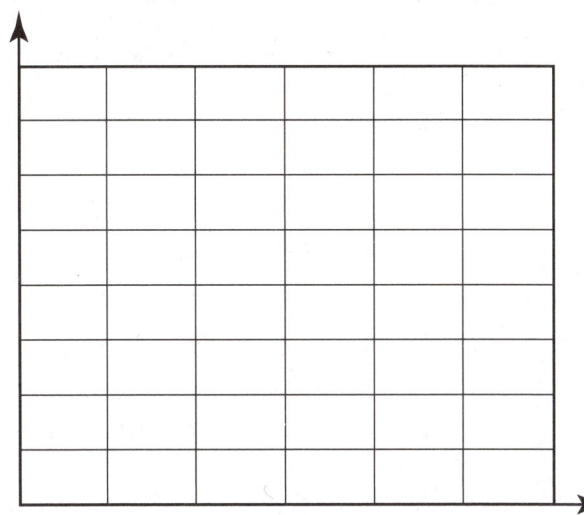

Lesson 4 Use the graph to answer each question.

③ Which story matches the graph? Circle A or B.

A. A shoe store sold 20 pairs of shoes on Monday. The store sold no shoes on Tuesday because it was closed. On Wednesday and Thursday, 30 pairs of shoes were sold.

B. A hot air balloon rose to 250 feet. It stayed there for a while and then rose to 500 feet. After a little while longer, the balloon began its descent.

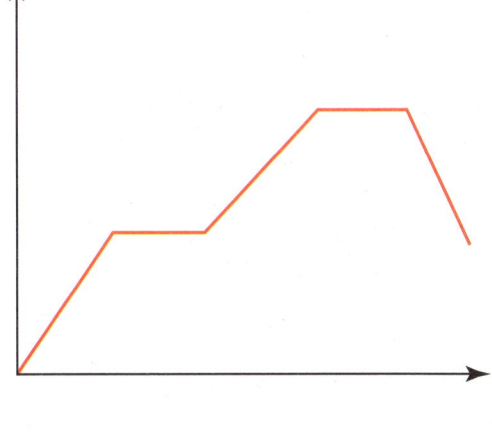

Reflect
What label would you put on the horizontal axis? What label would you put on the vertical axis?

_____

Week 4: Variables and Equality

Lesson 1

Key Idea

An **equation** is a number sentence which states that two mathematical expressions are equal.

$$2 + 3 = 5 \qquad 11 - 4 = 7 \qquad 6 - 2 = 3 + 1$$

Sometimes equations have unknown values. You can show unknown values with pictures, boxes, or letters.

$$4 + \square = 12 \qquad b - 9 = 5$$

Try This

Find the unknown value in each equation. Substitute values into the equation until you have a true number sentence.

1. $\square + 6 = 8$

   What is $\square$?

2. $\triangle + 1 = 7$

   What is $\triangle$?

3. $4 + \triangle = 8$

   What is $\triangle$?

4. $5 - \bigcirc = 2$

   What is $\bigcirc$?

5. $a + 8 = 10$

   $a = $ _____

6. $12 - 7 = z$

   $z = $ _____

Practice

Find the unknown value in each equation. The same shapes represent the same value.

7. $\square + \square = 10$

   What is $\square$?

8. $\bigcirc + \bigcirc = 2$

   What is $\bigcirc$?

32 Number Patterns and Relationships • Week 4

⑨ △ + △ = 6

What is △?

⑩ ◯ + ◯ + ◯ = 15

What is ◯?

Find the unknown value in each equation.

⑪ 9 − ☐ = 3

What is ☐?

⑫ 8 − n = 6

n = _____

⑬ t − 7 = 2

t = _____

⑭ 14 + ◯ = 19

What is ◯?

⑮ x + 4 = 12

x = _____

⑯ 17 − y = 12

y = _____

Reflect

What values of ☐ and △ make a true number sentence? Is there more than one correct answer? Explain.

☐ − △ = 3

_____

_____

_____

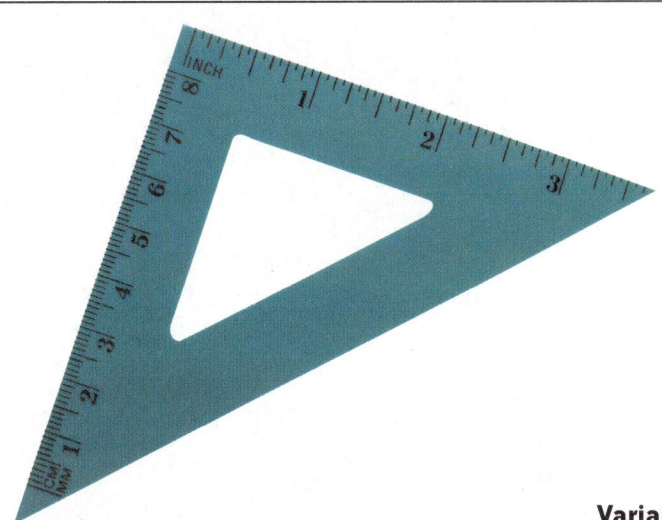

Variables and Equality • Lesson 1   33

Week 4 — Variables and Equality

Lesson 2

> **Key Idea**
> You can use the idea of weights to help solve equations.

Try This
Answer each question to find the weight of the toy car.

3 pounds

8 pounds

1 How much does the piggy bank weigh?

_____

2 How much do the piggy bank and toy car weigh altogether?

_____

3 Fill in the blanks below to help you find the weight of the toy car.

Piggy bank = 3

Piggy bank + toy car = 8

_____ + toy car = 8

_____ + _____ = 8

Toy car = _____

The toy car weighs _____.

Practice
Find each unknown weight. Write a number sentence to show your work.

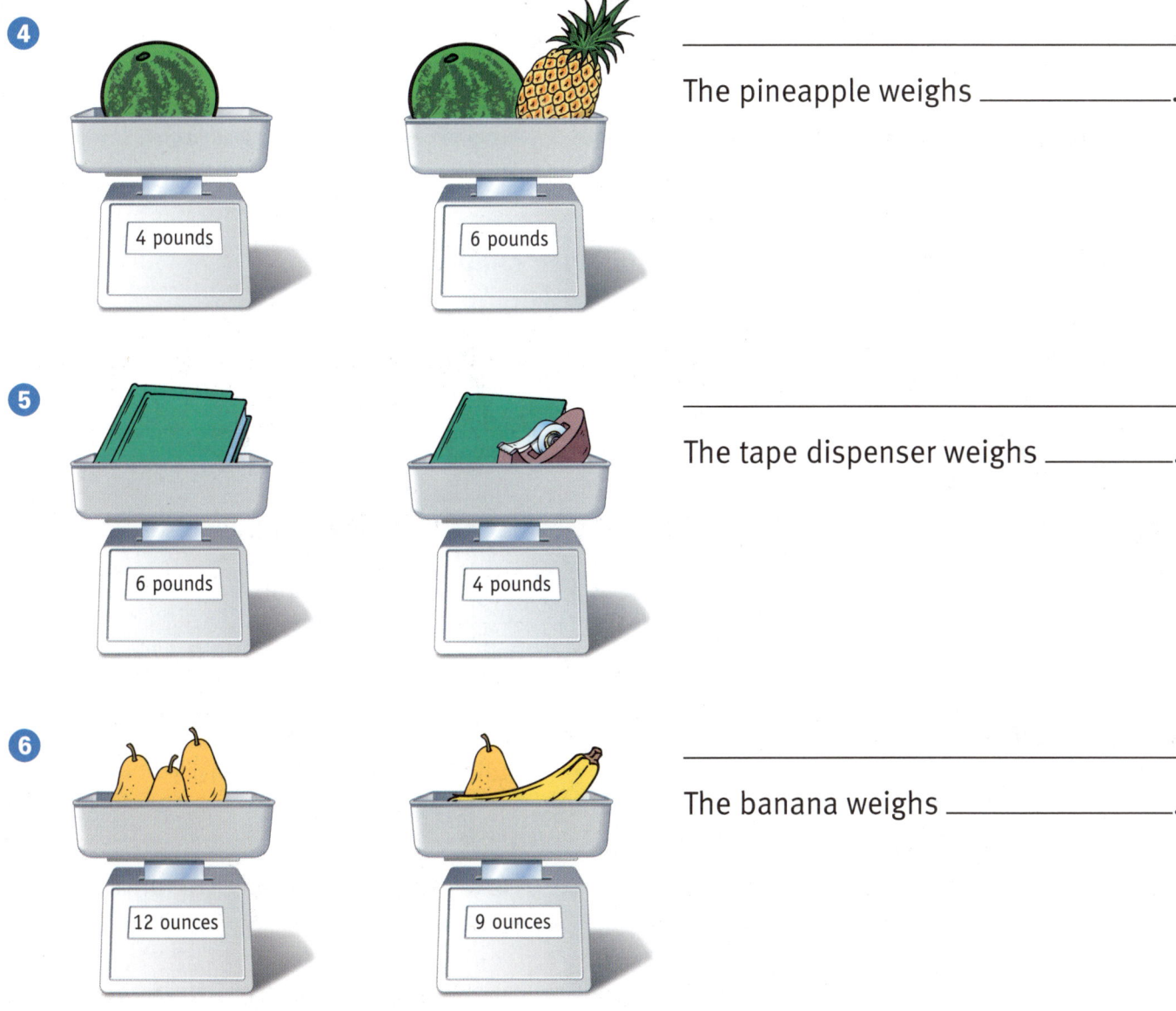

4 The pineapple weighs _____.

5 The tape dispenser weighs _____.

6 The banana weighs _____.

Reflect
How did you decide how much one pear weighs in Problem 6? Explain.

_____

_____

_____

Variables and Equality • Lesson 2   35

Week 4 — Variables and Equality

Lesson 3

> **Key Idea**
> Use reasoning to solve more challenging problems involving weights.

Try This
Answer each question to find the weight of each shape.

1. How much does the pyramid weigh?
   _____

2. How much do the pyramid and cylinder weigh altogether?
   _____

3. How much does the cylinder weigh?
   _____

4. How much do the cylinder and cube weigh altogether?
   _____

5. How much does the cube weigh?
   _____

Practice
Find each unknown weight.

6. The shoe weighs _____.
7. The basket weighs _____.
8. The plant weighs _____.

9. Each tennis ball weighs _____.
10. Each baseball weighs _____.
11. The volleyball weighs _____.

Reflect
Would you be able to find the weight of the volleyball in Problem 7 if you had only the first and third scales? Explain your answer.

_____

Variables and Equality • Lesson 3 37

Week 4 • Variables and Equality

Lesson 4

> **Key Idea**
> Balance scales can be used to help solve equations.
> When a scale is balanced, both sides are equal.

Try This
Use each balance scale to find two equal weights.

The weight of 1 orange is the same as the weight of _____.

The weight of 1 toy puppy is the same as the weight of
_____.

The weight of 1 box of crayons is the same as the weight of _____
_____.

The weight of 4 baseballs is the same as the weight of _____
_____.

Practice

Find each unknown weight. Draw your answer on the scale with the question mark.

Reflect

What part of a number sentence is represented by the balance scale? Explain.

_____

_____

_____

Variables and Equality • Lesson 4   **39**

Week 4 — Variables and Equality

Lesson 5 Review

This week you explored equality and unknown values in number sentences. You used shapes to represent missing numbers in an equation. You also related number sentences to weights and balance scales.

Lesson 1 Find the unknown value in each equation.

1. $c + 2 = 8$

   $c = $ _____

2. ☐ + ☐ + ☐ = 9

   What is ☐?

Lesson 2 Find each unknown weight. Write a number sentence to show your work.

3.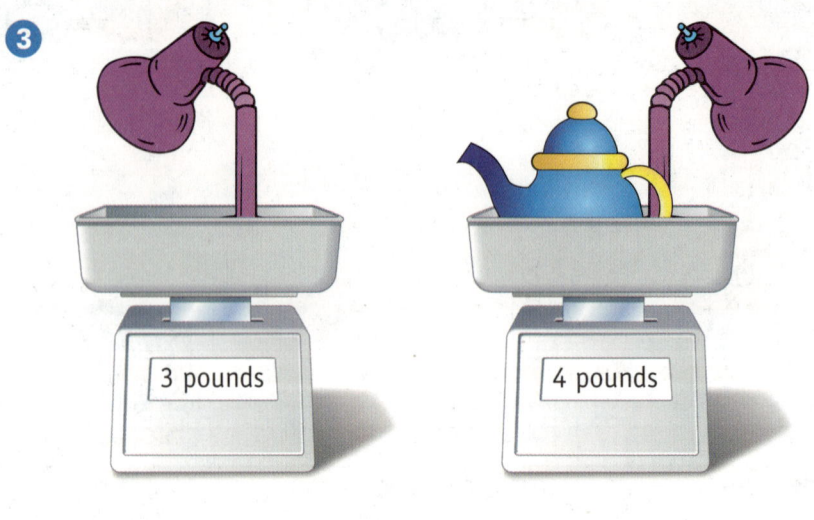

3 pounds   4 pounds

_____

The teapot weighs _____.

Reflect

Explain how to find the values of the unknowns in the number sentence.

☐ + ☐ + ☐ = 21

_____

_____

40   Number Patterns and Relationships • Week 4

Lesson 3

④

The knife weighs _____.

Lesson 4 Find the unknown weight. Draw your answer on the scale with the question mark.

⑤

Reflect
Use shapes from Problem 5 to balance each scale. Draw your answer on the scale with the question mark.

Variables and Equality • Lesson 5 Review

Week 1 Exploring Patterns

Practice

1. Draw the next set in the pattern.

Set 1   Set 2   Set 3   _____

2. Draw the missing set in the pattern.

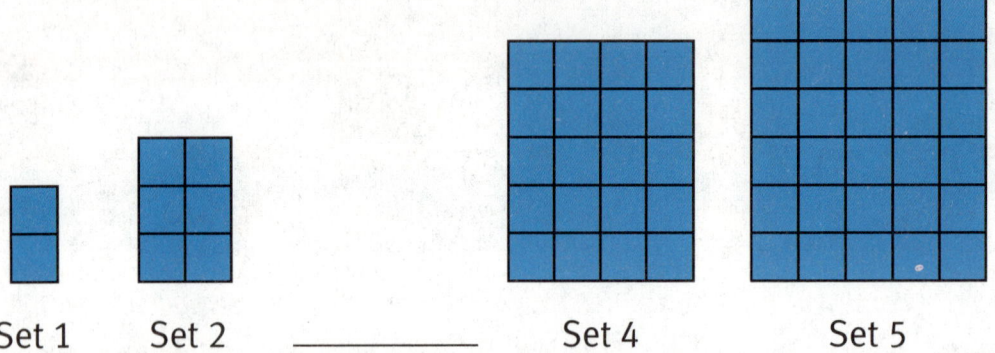

Set 1   Set 2   _____   Set 4   Set 5

3. Tell whether each pattern above is a same-step or a changing-step growing pattern.

_____

_____

4. Design a same-step or changing-step pattern. Show sets 1, 2, and 3, and describe the pattern.

42 Number Patterns and Relationships • Week 1 Practice

Patterns and Relationships

Practice

Complete the table for the pattern shown below.

1

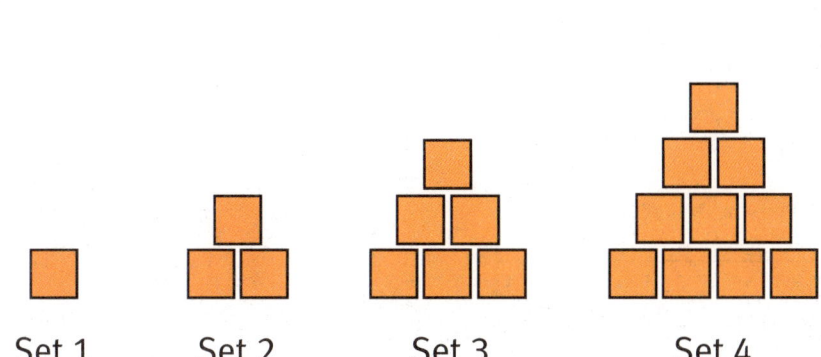

Set	Number of Squares

Complete each input/output table.

2 Candy bars cost 50¢ each.

Input (number of candy bars)	Output (total cost)

3 Jean can ride her bike 20 miles per hour.

Input (number of hours)	Output (number of miles)

Week 3 — Patterns and Graphs

Practice

1 The Booster Club sells gourmet cookies for $1.50 each. Complete the input/output table, and graph the pattern.

Input (number of cookies)	Output (total cost)

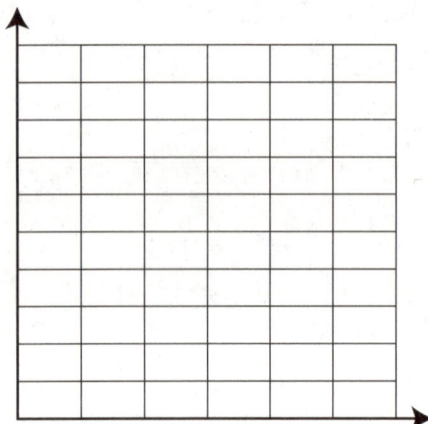

2 Describe the pattern that is shown in the graph.

___

3 How much would it cost to purchase 10 cookies?

___

4 What remains the same in the graph?

___

Week 4 — Variables and Equality

Practice

① $x - 5 = 2$
$x =$ _____

② $10 - y = 6$
$y =$ _____

③ $z + 5 = 15$
$z =$ _____

④ ☐ + ☐ + ☐ = 9
What is ☐?

⑤ $22 - \triangle = \triangle$
What is $\triangle$?

⑥ ◯ + ◯ = 12
What is ◯?

⑦ $\triangle + \triangle = 18$
What is $\triangle$?

⑧ $m - 6 = 2$
$m =$ _____

⑨ Explain how to find the values of the unknowns in the number sentence
☐ + ☐ + ☐ = 27

_____
_____
_____
_____

Unit 2 Workbook
Level F

NUMBER WORLDS

Number Patterns and Relationships

featuring Building Blocks Software

Author
Sharon Griffin
*Associate Professor of Education and
Adjunct Associate Professor of Psychology*
Clark University
Worcester, Massachusetts

Building Blocks Authors

Douglas H. Clements
*Professor of Early Childhood
and Mathematics Education*
University at Buffalo
State University of New York, New York

Julie Sarama
Associate Professor of Mathematics Education
University at Buffalo
State University of New York, New York

Contributing Writers
Sherry Booth, *Math Curriculum Developer,* Raleigh, North Carolina
Elizabeth Jimenez, *English Language Learner Consultant,* Pomona, California

Program Reviewers

Jean Delwiche
Almaden Country School
San Jose, California

Cheryl Glorioso
Santa Ana Unified School District
Santa Ana, California

Sharon LaPoint
School District of Indian River County
Vero Beach, Florida

Leigh Lidrbauch
Pasadena Independent School District
Pasadena, Texas

Dave Maresh
Morongo Unified School District
Yucca Valley, California

Mary Mayberry
Mon Valley Education Consortium, AIU 3
Clairton, Pennsylvania

Lauren Parente
Mountain Lakes School District
Mountain Lakes, New Jersey

Juan Regalado
Houston Independent School District
Houston, Texas

M. Kate Thiry
Dublin City School District
Dublin, Ohio

Susan C. Vohrer
Baltimore County Public Schools
Baltimore, Maryland

SRAonline.com

Copyright © 2007 SRA/McGraw-Hill.

All rights reserved. Except as permitted under the United States Copyright Act, no part of this publication may be reproduced or distributed in any form or by any means, or stored in a database or retrieval system, without the prior written permission of the publisher, unless otherwise indicated.

Printed in the United States of America.

Send all inquiries to:
SRA/McGraw-Hill
4400 Easton Commons
Columbus, OH 43219

R53180.01

7 8 9 QPE 12 11 10 09

Photo Credits
2–14 ©PhotoDisc/Getty Images, Inc.; **15** ©Stockbyte/Stockbyte; **16–23** ©PhotoDisc/Getty Images, Inc.; **24** ©Eyewire/Getty Images, Inc.; **25** ©Matt Meadows; **33** ©Stockbyte/Stockbyte

Contents

Number Patterns and Relationships

Week 1 Exploring Patterns 2

Week 2 Patterns and Relationships 12

Week 3 Patterns and Graphs 22

Week 4 Variables and Equality 32

Week 1 Practice 42

Week 2 Practice 43

Week 3 Practice 44

Week 4 Practice 45

Week 1

Exploring Patterns

Lesson 1

Key Idea

Same-step patterns are patterns in which the amount of change or growth is the same from one set to the next.

Ask the following questions to help you look for a pattern.

- What is changing from one set to the next?
- What stays the same from one set to the next?
- How would you create the next set?

Try This

Describe what changes in each pattern. Sketch and label the next figure in the pattern.

1

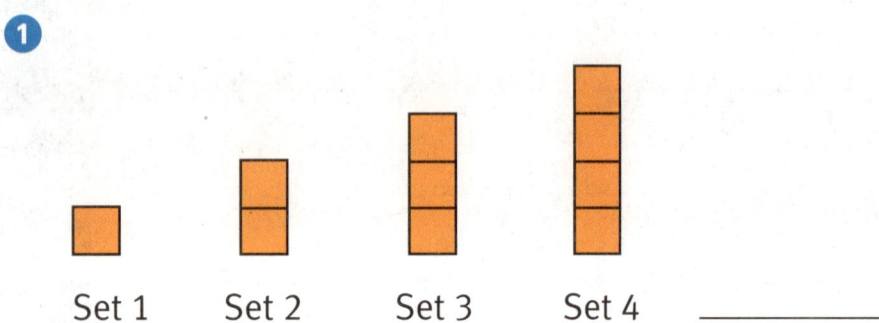

Set 1   Set 2   Set 3   Set 4   _____

_____

_____

2

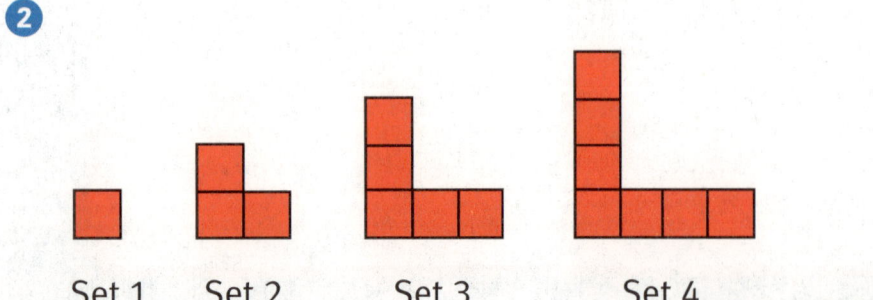

Set 1   Set 2   Set 3   Set 4   _____

_____

_____

2 Number Patterns and Relationships • Week 1

Practice
Sketch and label the next two terms in each pattern.

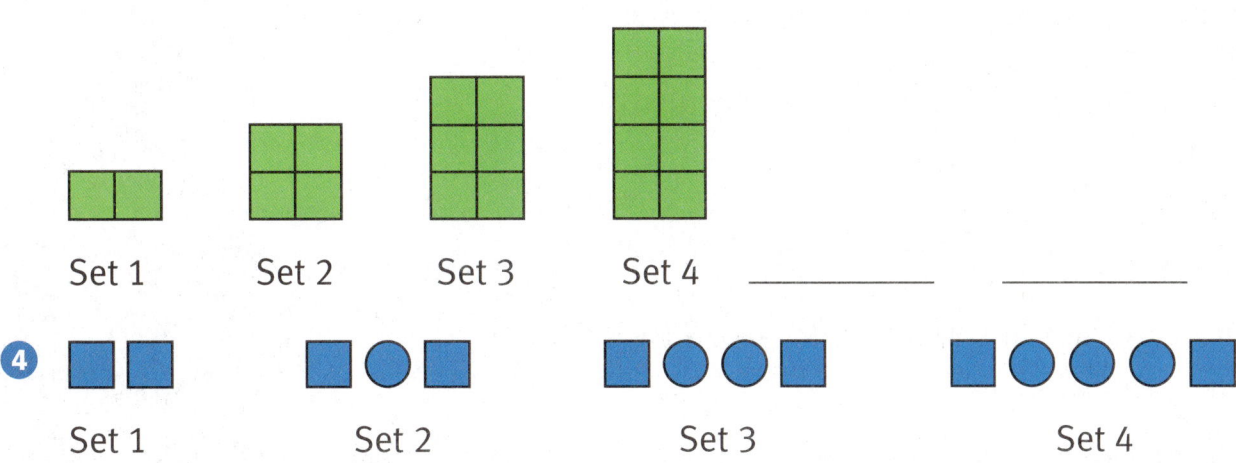

Reflect
Mandy had 5 pennies in her piggy bank on Monday. Each day she put 10 more pennies into the bank. How much money will she have in the piggy bank on Friday? Explain.

Monday	Tuesday	Wednesday	Thursday	Friday
5¢	15¢	25¢	35¢	_____

_____
_____
_____
_____

Exploring Patterns • Lesson 1

Week 1

Exploring Patterns

Lesson 2

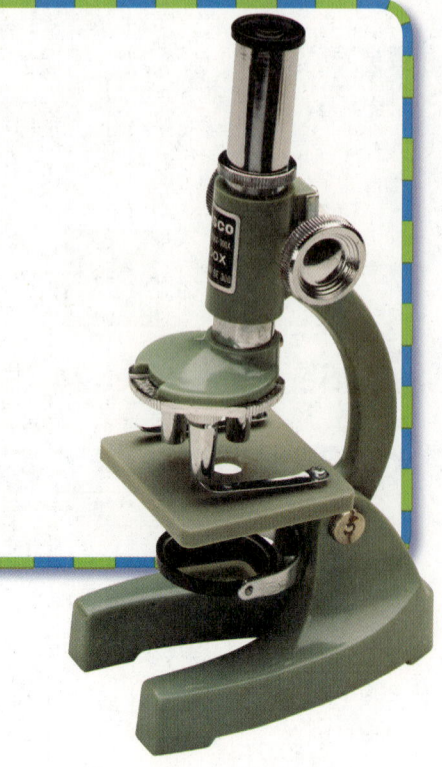

Key Idea

Changing-step patterns are patterns in which the amount of change or growth changes in a regular and predictable way from one set to the next.

Ask the following questions to help you identify a changing-step pattern:

- What changes from one set to the next?
- What stays the same from one set to the next?
- How would you create the next set?

Try This

Describe the changes in each pattern. Then sketch and label the next set in the pattern.

1

   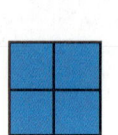   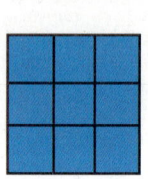   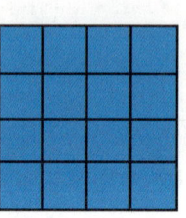

Set 1   Set 2   Set 3   Set 4   _____

_____

_____

2

Set 1   Set 2   Set 3   _____

_____

_____

4 Number Patterns and Relationships • Week 1

Practice
Sketch and label the next term in the pattern.

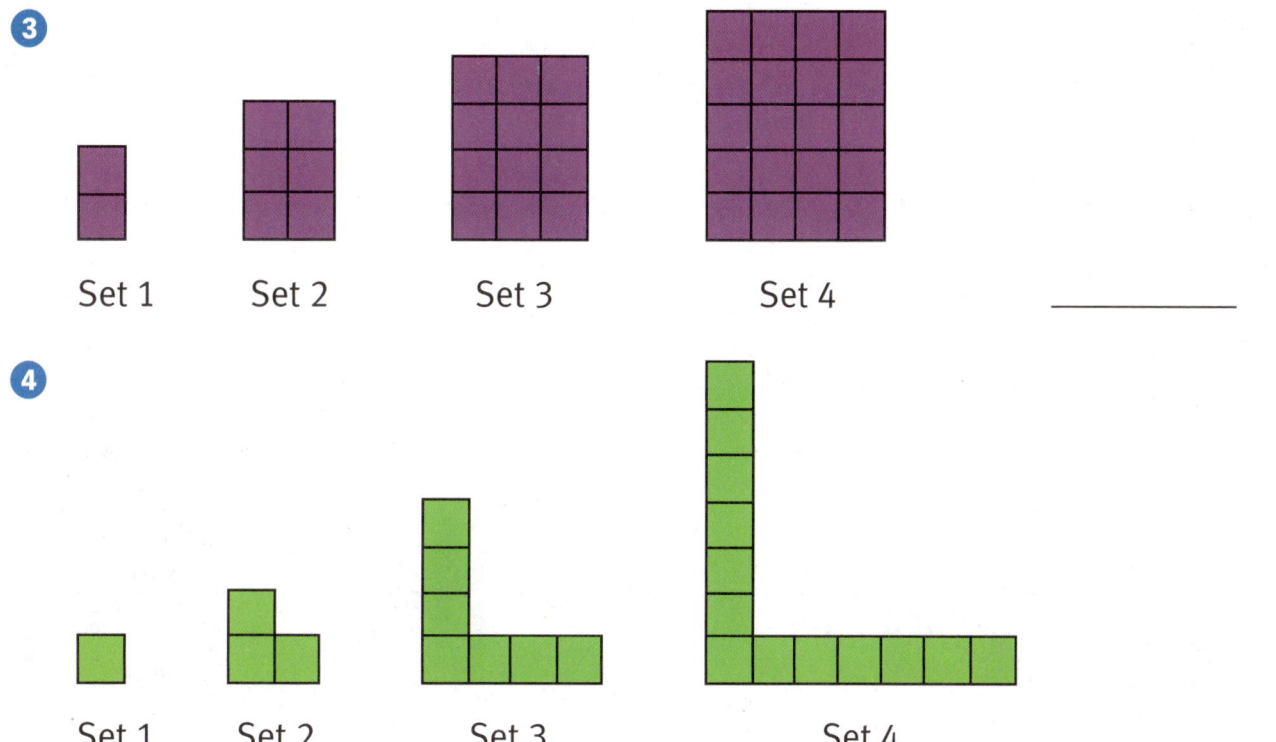

3 Set 1   Set 2   Set 3   Set 4   _____

4 Set 1   Set 2   Set 3   Set 4   _____

Reflect
Look for a pattern in the table. If the pattern continues, how many chin-ups will Jarrod do on Friday? Explain your answer.

Jarrod's Chin-up Chart				
Monday	Tuesday	Wednesday	Thursday	Friday
8 chin-ups	10 chin-ups	13 chin-ups	17 chin-ups	

_____
_____
_____

Week 1

Exploring Patterns

Lesson 3

> **Key Idea**
> You can use clues from the sets given in a pattern to identify missing sets.

Try This
Look for the changes in each pattern. Then draw and label the missing set.

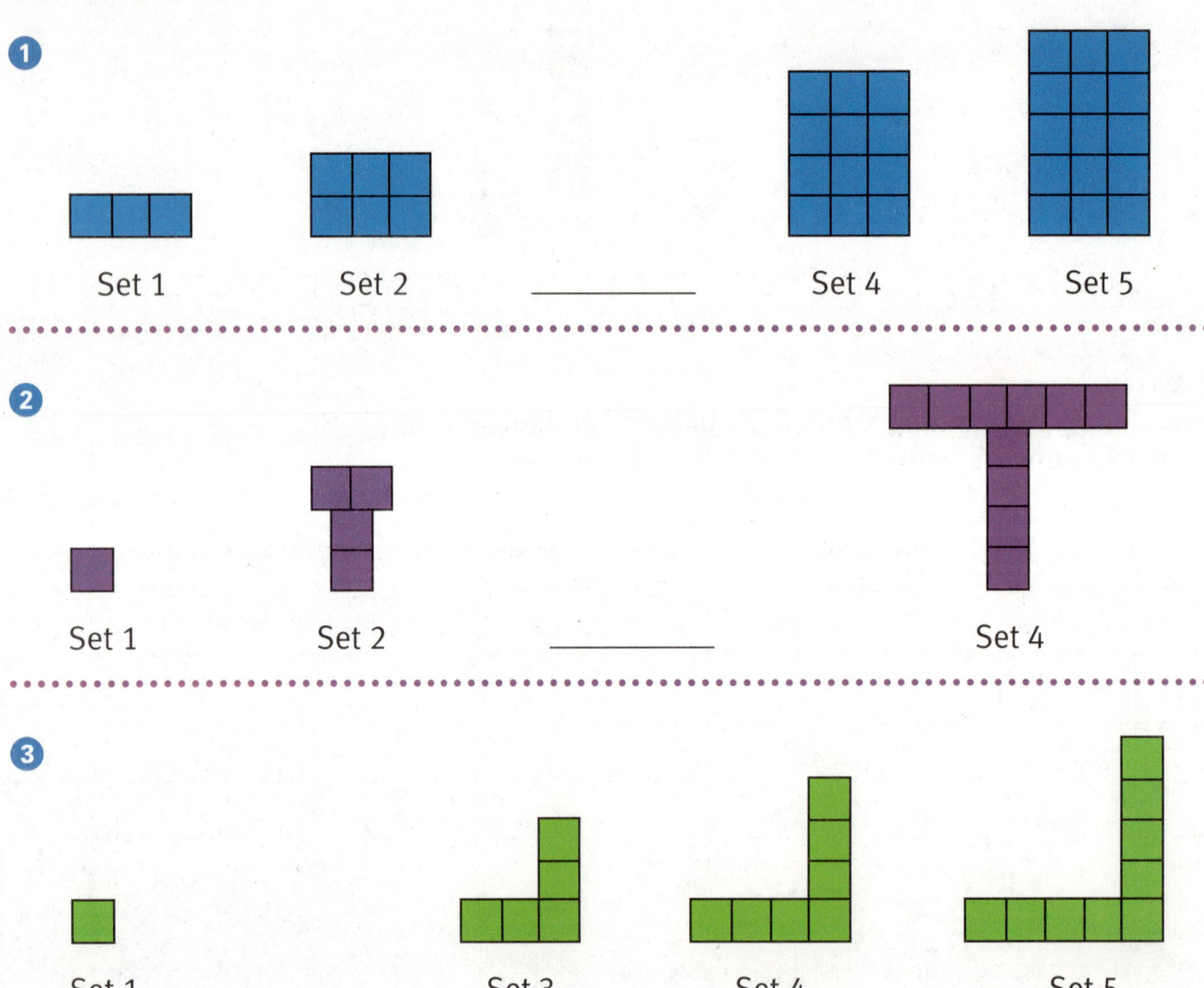

6 Number Patterns and Relationships • Week 1

Practice
Look for the changes in each pattern. Tell whether it is a same-step growing pattern or a changing-step growing pattern. Then draw and label the missing set.

4.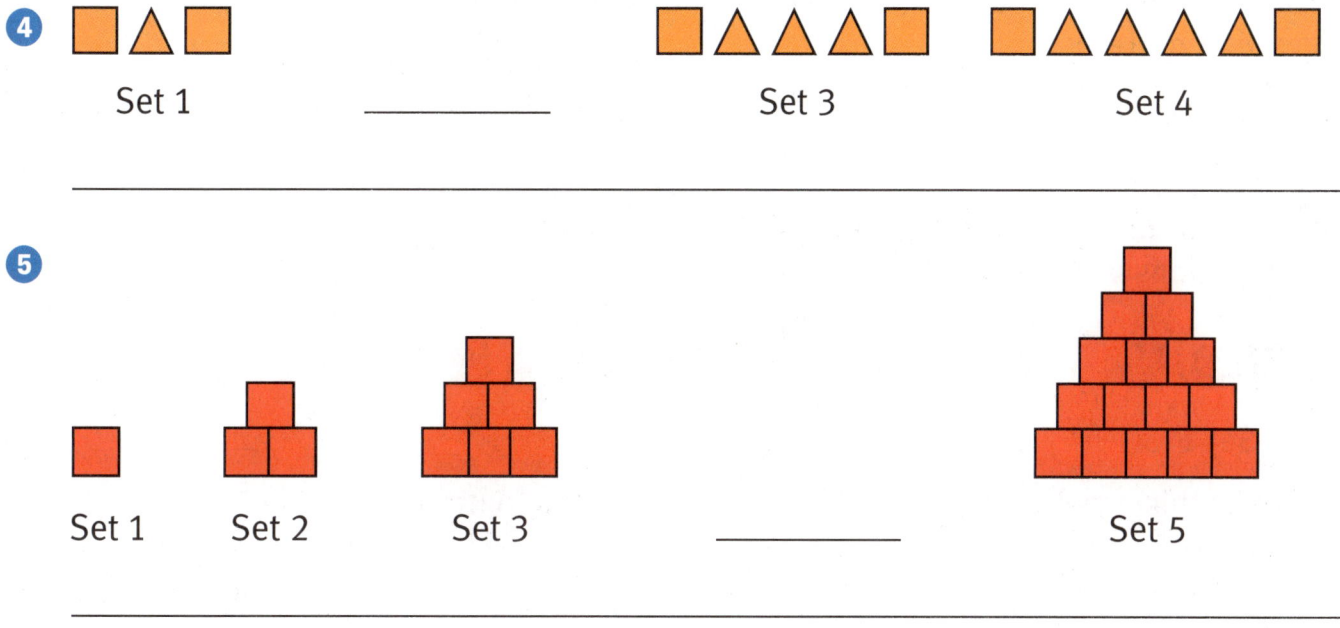

5.

Reflect
Do you think it is easier to find a missing set in a same-step or a changing-step growing pattern? Explain.

_____
_____
_____
_____

Week 1 — Exploring Patterns

Lesson 4

Key Idea
You can use pattern blocks to create your own growing patterns.

Try This
Draw and label the next set for each pattern. Use the pattern to answer the questions.

1

   Set 1   Set 2   Set 3

a. Is this a same-step or a changing-step growing pattern?

_____

b. How is the pattern changing from set to set?

_____

2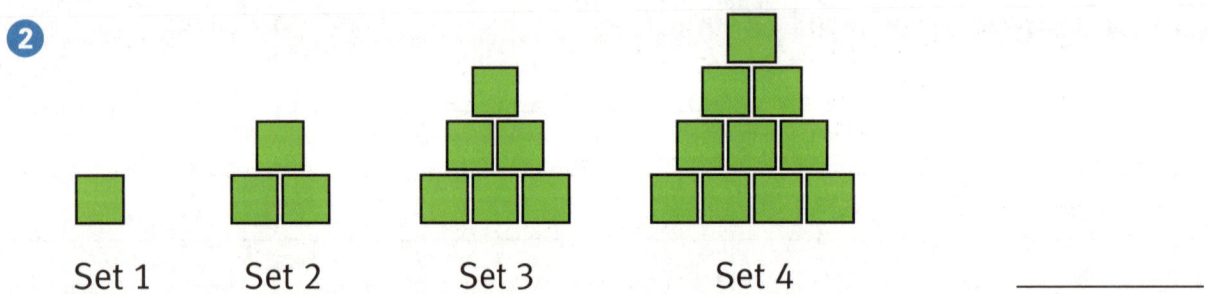

Set 1   Set 2   Set 3   Set 4

a. Is this a same-step or a changing-step growing pattern?

_____

b. How is the pattern changing from set to set?

_____

_____

8 Number Patterns and Relationships • Week 1

Practice
Draw your own pattern in the space below. Exchange your pattern with a partner, and have your partner answer each question.

❸ Is this a same-step or a changing-step growing pattern?
_____

❹ How is the pattern changing from set to set?
_____

❺ What would the next set of the pattern look like? Describe it.
_____

Reflect
Use different shapes to create the first few sets of a pattern. Explain how the pattern changes from set to set.

_____

_____

Exploring Patterns • Lesson 4

Week 1 Exploring Patterns

Lesson 5 Review

This week you explored patterns. You examined same-step patterns and changing-step patterns.

Lesson 1 Sketch and label the next set in the pattern.

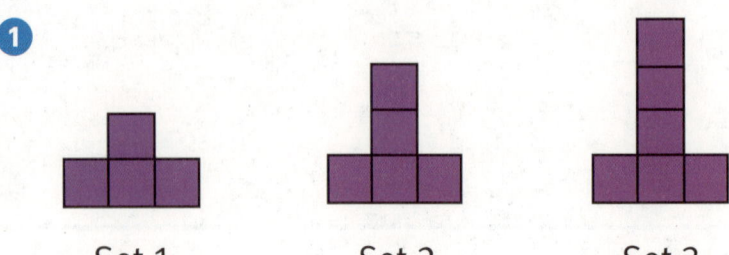

Lesson 2 Sketch and label the next set in the pattern.

Reflect
Tell whether each pattern above is a same-step or a changing-step growing pattern.

_____
_____
_____

10 Number Patterns and Relationships • Week 1

Lesson 3 Sketch and label the missing set in the pattern.

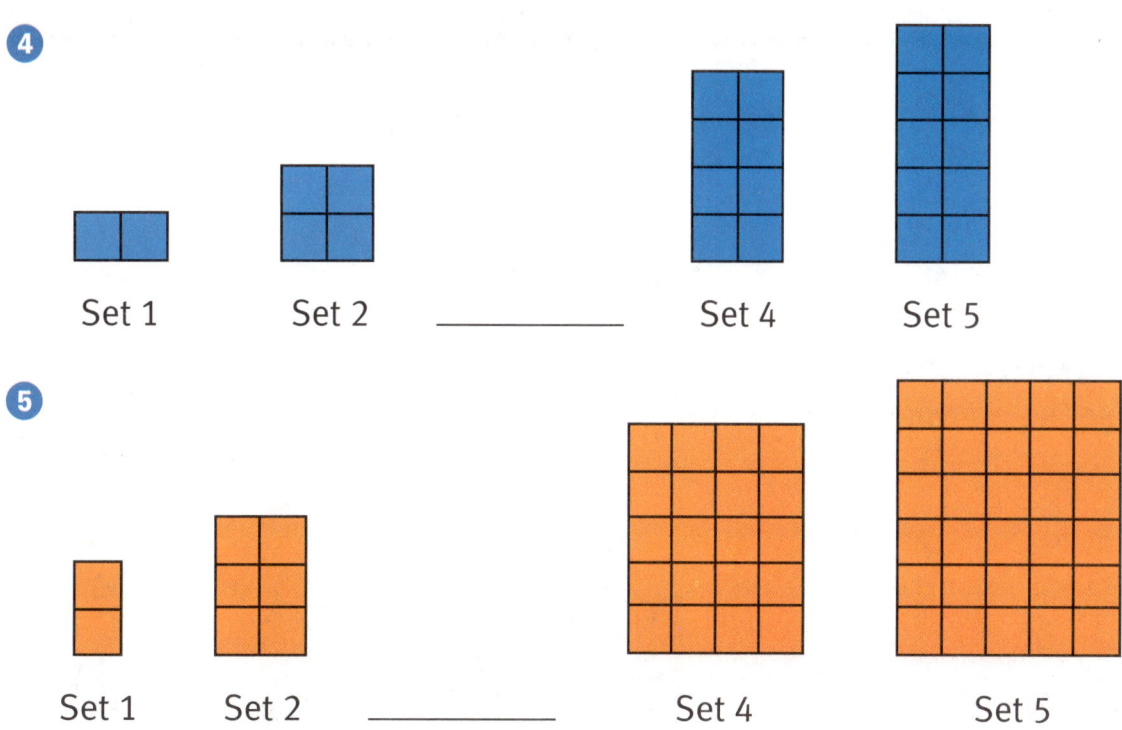

④ Set 1   Set 2   _____   Set 4   Set 5

⑤ Set 1   Set 2   _____   Set 4   Set 5

Lesson 4

⑥ Use pattern blocks to design Set 1 of a pattern. Then show Sets 2 and 3.

Reflect
Describe how to find the missing set. Then draw and label the missing set.

Set 1   Set 2   _____   Set 4

_____

_____

_____

_____

Exploring Patterns • Lesson 5 Review 11

Week 2

Patterns and Relationships

Lesson 1

> **Key Idea**
> You can use numbers to represent the growth in patterns.

Try This
Use the growing pattern below to answer each question.

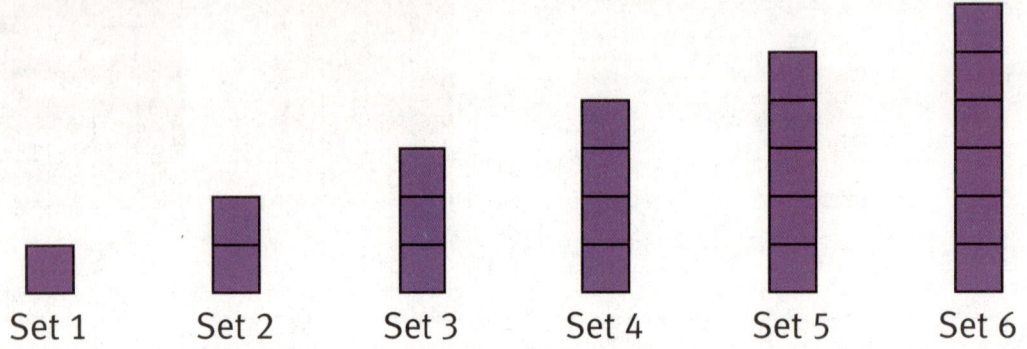

1. Use the pattern to complete the table below.

Set	Number of Squares
1	
2	
3	
4	
5	
6	

2. What patterns do you notice in the table between the set number and the number of squares?

_____

_____

3. Use the pattern from Problem 2 to predict the number of squares in the tenth set without building or drawing all the sets up to the tenth.

_____

12 Number Patterns and Relationships • Week 2

Practice
Use the pattern to complete the table.

4

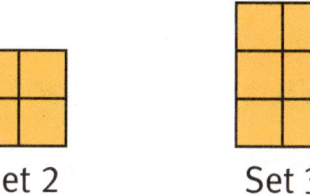
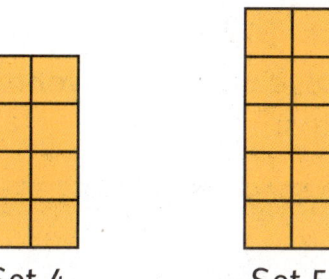

Set 1   Set 2   Set 3   Set 4   Set 5

Set	Number of Squares
1	
2	
3	
4	
5	

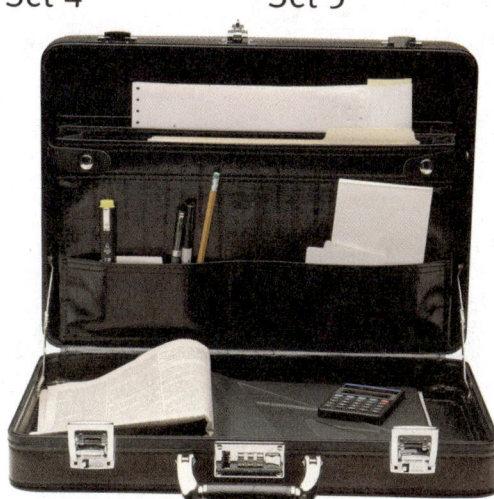

5 What patterns do you notice in the table above between the set number and the number of squares?

_____

6 Use the pattern from Problem 5 to predict the number of squares in the tenth set without building or drawing all of the sets up to the tenth.

_____

Reflect
When you use a pattern table to help you create a pattern with shapes, what is the smallest number of sets you need to figure out the pattern? Explain.

_____

_____

Week 2 — Patterns and Relationships

Lesson 2

Key Idea
You can use numbers to represent the growth in patterns.

Try This
Use the growing pattern below to answer each question.

Set 1   Set 2   Set 3   Set 4   Set 5

1. Is this a same-step or a changing-step growth pattern?
   _____

2. Describe the growth shown in the pattern.
   _____

3. How many squares were used to create each set?
   _____

4. Use your answers to complete the table below.

Set	Number of Squares
1	
2	
3	
4	
5	

14   Number Patterns and Relationships • Week 2

Practice
Use the pattern to complete the table.

5

  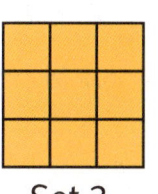  

Set 1   Set 2   Set 3   Set 4   Set 5

Set	Number of Squares
1	
2	
3	
4	
5	

6 What patterns do you notice in the table above between the set number and the number of squares?

___

7 Use the pattern to predict the number of squares in the tenth set without building or drawing all of the sets up to the tenth.

___

Reflect
What do you notice about the table for a changing-step pattern? How does it compare with the table for a same-step pattern?

___
___
___
___

Patterns and Relationships • Lesson 2

Week 2 — Patterns and Relationships

Lesson 3

> **Key Idea**
>
> The pattern tables are examples of input/output tables.
>
> For each input value (the set number), there is a certain output value (number of squares).
>
> Input/output tables can also be used to model real-world situations.

Try This

Thomas mows lawns in his neighborhood to earn money. He earns $8 for each lawn he mows. Use the input/output table to answer each question.

Input (lawns mowed)	Output (money earned)
1	$8
2	$16
3	$24
4	$32
5	$40

1 How much money would Thomas earn if he mowed 6 lawns? Explain how you found your answer.

_____

_____

_____

2 Write a mathematical rule for determining the amount of money Thomas will earn for any given number of lawns mowed.

_____

16 Number Patterns and Relationships • Week 2

Practice
Complete the table and answer each question.

❸ Movie tickets cost $6 each. Complete the input/output table.

Input (number of tickets)	Output (total cost)
1	$6
2	
3	
4	
5	

❹ How much would it cost to buy 4 movie tickets?

_____

❺ Write a mathematical rule for determining the total cost of tickets for any given number of tickets.

_____

_____

❻ For every hour Lisa drives, she uses 2 gallons of gasoline. Her gas tank holds 18 gallons when it is full. Complete the input/output table.

Input (hours of driving)	Output (gas remaining in her tank)
1	16 gallons
2	14 gallons
3	
4	
5	

❼ How much gasoline is in Lisa's tank after 5 hours of driving?

_____

❽ Write a mathematical rule for determining the amount of gas remaining for any given number of hours driven.

_____

_____

Reflect
What is different about the lawn mowing input/output table and the gasoline input/output table?

_____

_____

_____

Patterns and Relationships • Lesson 3   **17**

Week 2 — Lesson 4

Patterns and Relationships

Key Idea
You can use input/output tables to help you make choices.

Try This

Anna's neighbors have hired her to pet sit their dog for seven days. They have offered two different options for being paid.

- **Option 1:** Anna receives $10 for the first day and an additional $2 per day after the first day.

- **Option 2:** Anna receives $1 for the first day. Every day after the first day she receives an additional amount that is $1 more than the previous day.

1 Complete the input/output tables for each option.

Option 1	
Day	Total Amount Earned
1	$10
2	$12
3	
4	
5	
6	
7	

Option 2	
Day	Total Amount Earned
1	$1
2	$3
3	$6
4	
5	
6	
7	

2 If Anna chooses Option 1, how much money will she be paid? If Anna chooses Option 2, how much money will she be paid?

Practice

Jim was hired to do yard work for his neighbor. The neighbor expects the work to last 5 days, but it could last 7 days. Payment options are as follows:

- **Option 1:** Jim receives $12 for the first day and $2 per day after the first day.

- **Option 2:** Jim receives $2 for the first two days. Every day thereafter he receives an amount that is $1 more than the previous day.

Complete the tables to find Jim's total earnings for the week for each option.

❸ Complete the table for each option.

Option 1				Option 2		
Day	Amount Earned for the Day	Total Earnings		Day	Amount Earned for the Day	Total Earnings
1	$12			1	$2	
2				2		
3				3		
4				4		
5				5		
6				6		
7				7		

❹ How much will Jim earn for 7 days if he chooses Option 1? _____

❺ How much will Jim earn for 7 days if he chooses Option 2? _____

❻ If the job is for only 5 days, which option will pay better? _____

Reflect

What kind of growth pattern is shown by Option 1? What kind of growth pattern is represented by Option 2?

Patterns and Relationships • Lesson 4

Week 2 Patterns and Relationships

Lesson 5 Review

This week you explored patterns and relationships. You looked at how visual patterns can be related to number patterns. You also learned about input/output tables and solving problems.

Lessons 1 and 2

Complete the table for the pattern shown below.

①

Set 1   Set 2   Set 3   Set 4

Set	Number of Cubes
1	
2	
3	
4	

Reflect
How is the pattern changing?

_____

Lesson 3 Complete the input/output table.

❷ The bookstore sells pencils for 15¢ each.

Input (number of pencils)	Output (total cost)
1	15¢
2	
3	
4	

Lesson 4 A bathtub holds 60 gallons of water. When the drain plug is pulled, 12 gallons drain from the tub each minute.

❸ How long does it take for the tub to fully drain?

_____

Input (number of minutes)	Output (water remaining in the tub)
0	60 gallons
1	
2	
3	
4	
5	

Reflect
How many gallons of water are left in the tub 1 minute after the plug is pulled? How many gallons of water are left in the tub 3 minutes after the plug is pulled? Is this an example of same-step pattern or a changing-step pattern?

_____

_____

_____

Patterns and Relationships • Lesson 5 Review

Week 3 — Patterns and Graphs

Lesson 1

> **Key Idea**
> Patterns can be represented with pictures, rules, and tables. They can also be represented with graphs.

Try This

Below is a graph that shows how far Lisa can drive, depending on the number of gallons of gasoline in the car's tank.

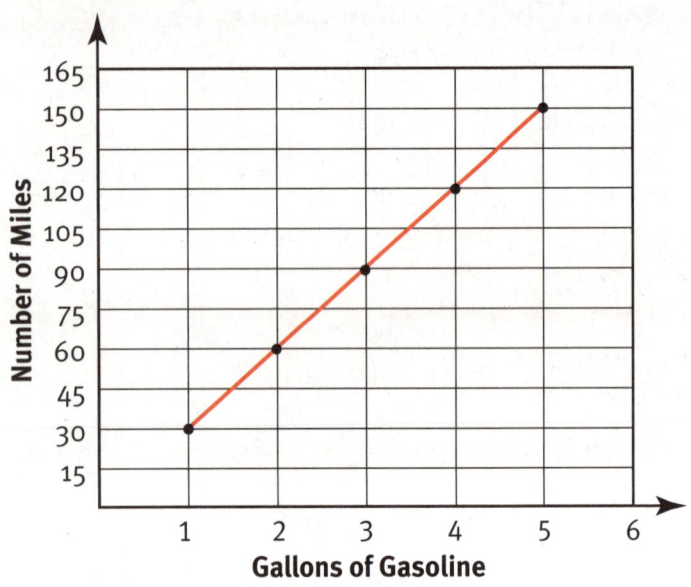

1. Which axis represents the number of miles Lisa can drive?

   _____

2. Which axis represents the amount of gasoline in Lisa's car?

   _____

3. As the number of gallons of gasoline increases, what happens to the distance that Lisa can drive? Is this increase a same-step increasing pattern or a changing-step pattern?

   _____
   _____

4. Describe the pattern shown in the graph.

   _____
   _____
   _____

Practice
The graph shows the money Thomas made mowing lawns.

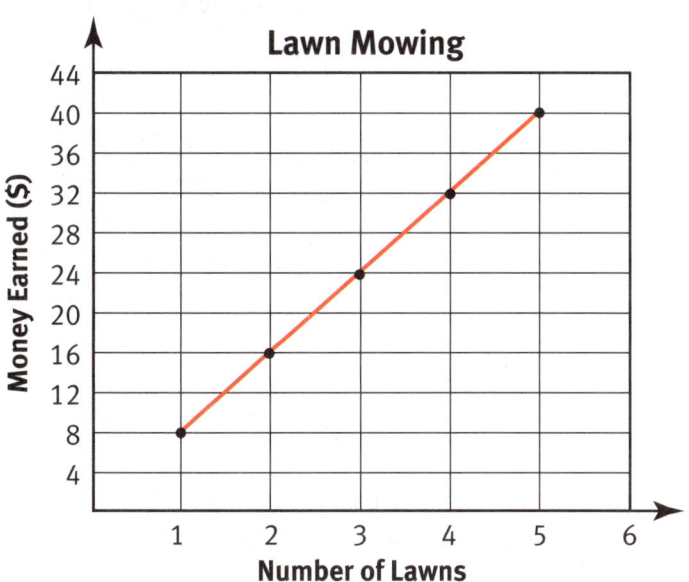

5 The data points are connected to help you see the trend of the data. As the number of lawns mowed increases, what happens to the amount of money earned?

_____

6 By connecting the data points, you are showing that the data is continuous. Should these data points be connected? Explain your answer.

_____
_____
_____
_____

Reflect
Create an input/output table, using the information above.

Week 3 — Patterns and Graphs

Lesson 2

> ## Key Idea
> When creating a graph, be sure to label the axes and give it a title.

Try This
Follow the steps to create a graph of the pattern.

Input (movie tickets)	Output (total cost)
1	$6
2	$12
3	$18
4	$24
5	$30

Step 1 Label the horizontal axis and the vertical axis.

Step 2 Plot a point for each pair of numbers in the table.

Step 3 Give your graph a title.

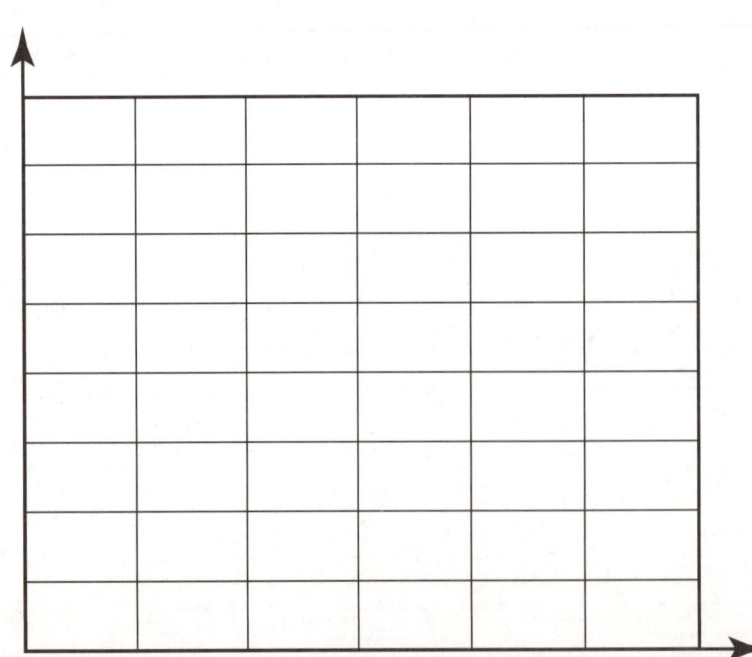

Practice

Mandy is babysitting for her neighbors. Graph the pattern shown in the input/output table.

Input (number of hours)	Output (money earned)
1	$5
2	$10
3	$15
4	$20
5	$25

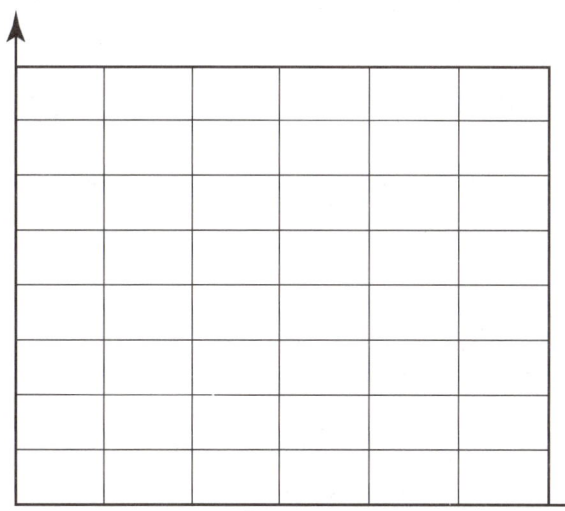

① Describe the pattern shown in the graph.

_____

② How much does Mandy earn for babysitting 6 hours?

_____

Reflect

Can you determine the rule for a pattern by just looking at the graph? Explain and give an example.

_____

_____

_____

Patterns and Graphs • Lesson 2

Week 3

Patterns and Graphs

Lesson 3

Key Idea
You can use graphs to compare two related patterns.

Try This
Create a graph for each input/output table.
Answer each question.

Milk Cartons Sold

Input (day)	Output (milk sold for the week)
Monday	25 cartons
Tuesday	50 cartons
Wednesday	75 cartons
Thursday	100 cartons
Friday	125 cartons

Milk Cartons Left

Input (day)	Output (milk cartons left in the cafeteria)
Monday	125 cartons
Tuesday	100 cartons
Wednesday	75 cartons
Thursday	50 cartons
Friday	25 cartons

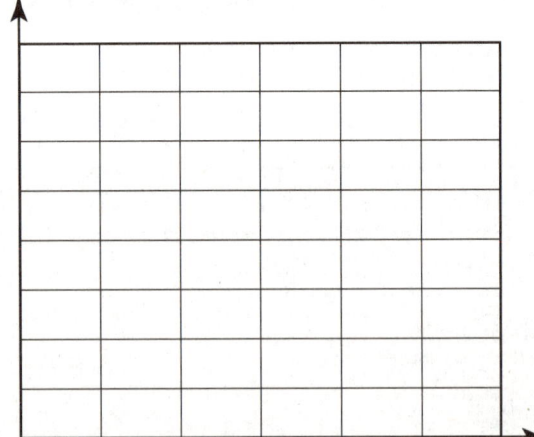
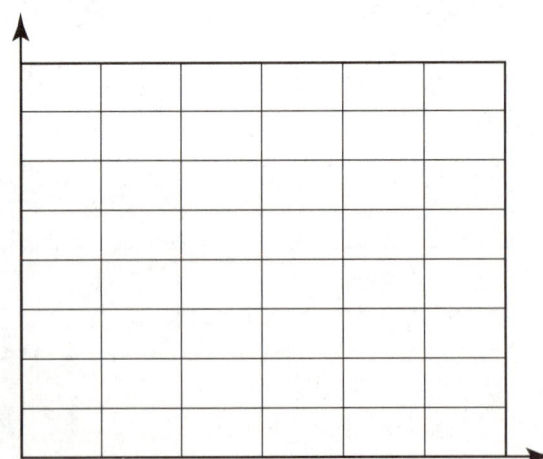

1 Which arrow line or axis represents the day of the week? Which represents the number of milk cartons left?

_____

_____

26 Number Patterns and Relationships • Week 3

Practice
Use your graphs from Try This to answer each question.

② Describe the pattern shown in the first graph.
_____

③ Describe the pattern shown in the second graph.
_____

④ Which of the graphs shows a growing pattern?
_____

⑤ Are these graphs same-step patterns or changing-step patterns?
_____

⑥ What stays the same in the first graph? What changes?
_____

⑦ What stays the same in the second graph? What changes?
_____

⑧ How are the two graphs related?
_____
_____

Reflect
Can you show both patterns on the same graph? Explain.
_____
_____
_____

Patterns and Graphs • Lesson 3

Week 3

Patterns and Graphs

Lesson 4

Key Idea
You can use graphs to tell a story or make an informed decision.

Try This
Choose the story that belongs with each graph.

Story A Melissa rode her bike for 40 minutes. The table shows the distance she traveled.

Time	10 minutes	20 minutes	30 minutes	40 minutes
Distance	3 miles	6 miles	9 miles	12 miles

Story B Mrs. Swanson walked her dog for 40 minutes. The table shows the number of blocks she covered.

Time	10 minutes	20 minutes	30 minutes	40 minutes
Distance	6 blocks	12 blocks	18 blocks	24 blocks

1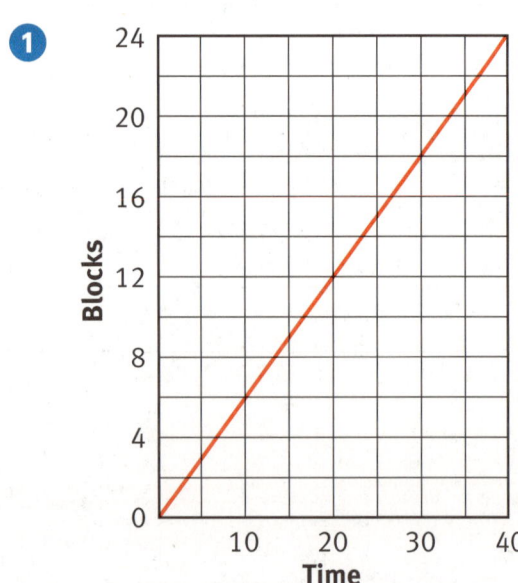

2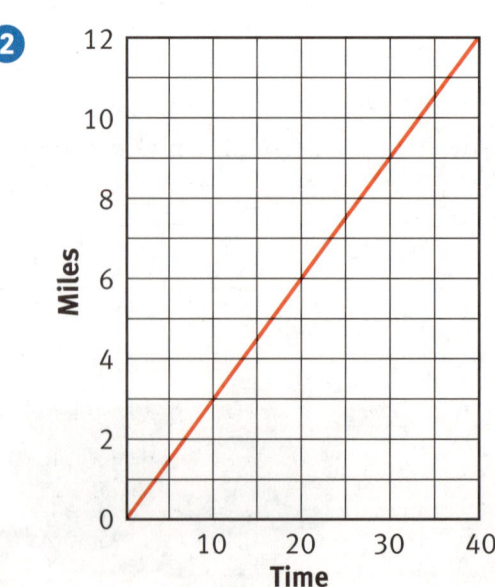

_____   _____

28 Number Patterns and Relationships • Week 3

Practice
Create a graph for the data in the table.

③
Number of Books	1	2	4	6
New Vocabulary Words	2	4	8	12

Reflect
Create a table that compares the outside temperature throughout the morning and afternoon of a winter day. Graph the data in the table.

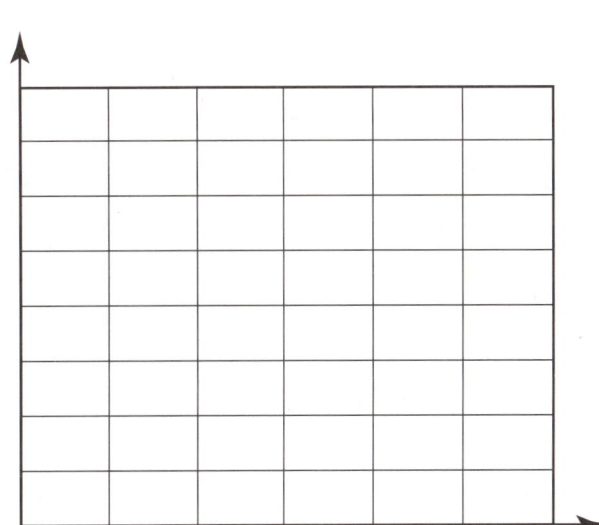

Patterns and Graphs • Lesson 4   29

Week 3

Patterns and Graphs

Lesson 5 Review

This week you explored how patterns look in graphs. You used input/output tables and stories to create graphs. You also used graphs to answer questions about the pattern and data.

Lessons 1 and 2

The bookstore sells school sweatshirts for $10 each. Graph the pattern shown in the input/output table.

Input (number of sweatshirts)	Output (total cost)
1	$10
2	$20
3	$30
4	$40
5	$50

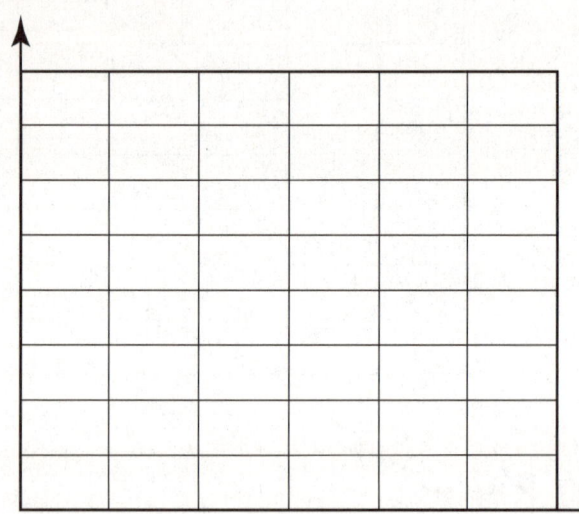

① Describe the pattern that is shown in the graph.

_____

_____

② How much would it cost to purchase eight sweatshirts?

_____

Reflect
What is staying the same in the graph?

_____

30   Number Patterns and Relationships • Week 3

Lesson 3 Graph the pattern shown in the input/output table.

Input (number of cars)	Output (total passengers)
1	4
2	8
3	12
4	16
5	20

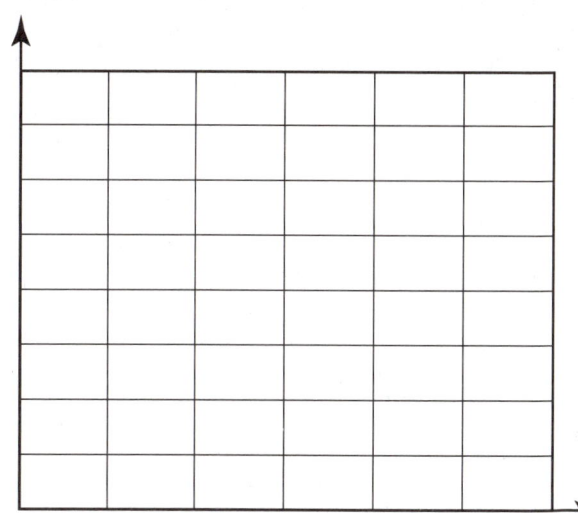

Lesson 4 Use the graph to answer each question.

③ Which story matches the graph? Circle A or B.

A. A shoe store sold 20 pairs of shoes on Monday. The store sold no shoes on Tuesday because it was closed. On Wednesday and Thursday, 30 pairs of shoes were sold.

B. A hot air balloon rose to 250 feet. It stayed there for a while and then rose to 500 feet. After a little while longer, the balloon began its descent.

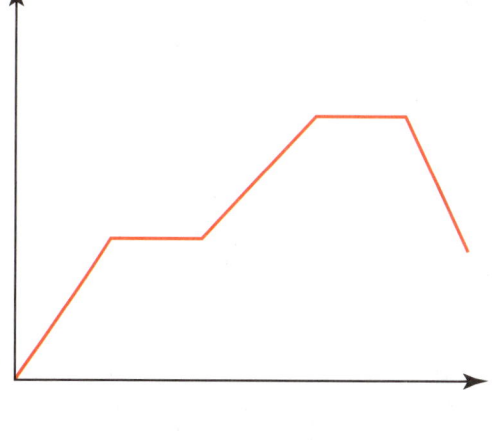

Reflect
What label would you put on the horizontal axis? What label would you put on the vertical axis?

Patterns and Graphs • Lesson 5 Review

Week 4 — Variables and Equality

Lesson 1

Key Idea

An **equation** is a number sentence which states that two mathematical expressions are equal.

$2 + 3 = 5$ $\qquad$ $11 - 4 = 7$ $\qquad$ $6 - 2 = 3 + 1$

Sometimes equations have unknown values. You can show unknown values with pictures, boxes, or letters.

$4 + \square = 12$ $\qquad$ $b - 9 = 5$

Try This

Find the unknown value in each equation. Substitute values into the equation until you have a true number sentence.

1. $\square + 6 = 8$

   What is $\square$?

2. $\triangle + 1 = 7$

   What is $\triangle$?

3. $4 + \triangle = 8$

   What is $\triangle$?

4. $5 - \bigcirc = 2$

   What is $\bigcirc$?

5. $a + 8 = 10$

   $a = $ _____

6. $12 - 7 = z$

   $z = $ _____

Practice

Find the unknown value in each equation. The same shapes represent the same value.

7. $\square + \square = 10$

   What is $\square$?

8. $\bigcirc + \bigcirc = 2$

   What is $\bigcirc$?

9 △ + △ = 6

What is △?

10 ◯ + ◯ + ◯ = 15

What is ◯?

Find the unknown value in each equation.

11 9 − ☐ = 3

What is ☐?

12 8 − n = 6

n = _____

13 t − 7 = 2

t = _____

14 14 + ◯ = 19

What is ◯?

15 x + 4 = 12

x = _____

16 17 − y = 12

y = _____

Reflect

What values of ☐ and △ make a true number sentence? Is there more than one correct answer? Explain.

☐ − △ = 3

_____

_____

_____

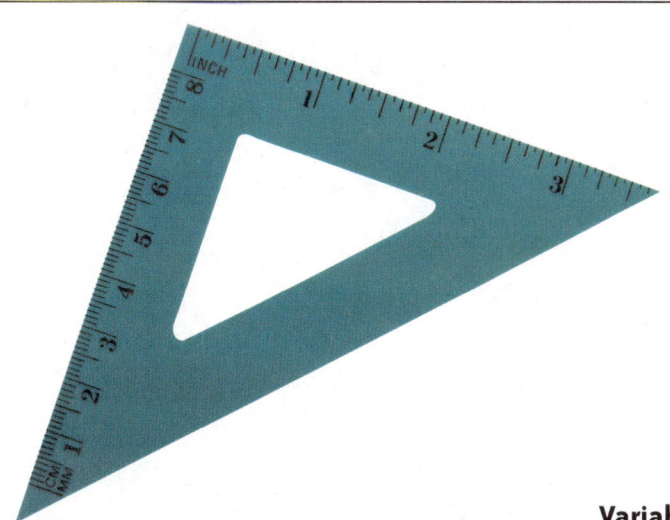

Variables and Equality • Lesson 1   33

Week 4: Variables and Equality

Lesson 2

Key Idea
You can use the idea of weights to help solve equations.

Try This
Answer each question to find the weight of the toy car.

3 pounds

8 pounds

1 How much does the piggy bank weigh?

_____

2 How much do the piggy bank and toy car weigh altogether?

_____

3 Fill in the blanks below to help you find the weight of the toy car.

Piggy bank = 3

Piggy bank + toy car = 8

_____ + toy car = 8

_____ + _____ = 8

Toy car = _____

The toy car weighs _____.

34 Number Patterns and Relationships • Week 4

Practice

Find each unknown weight. Write a number sentence to show your work.

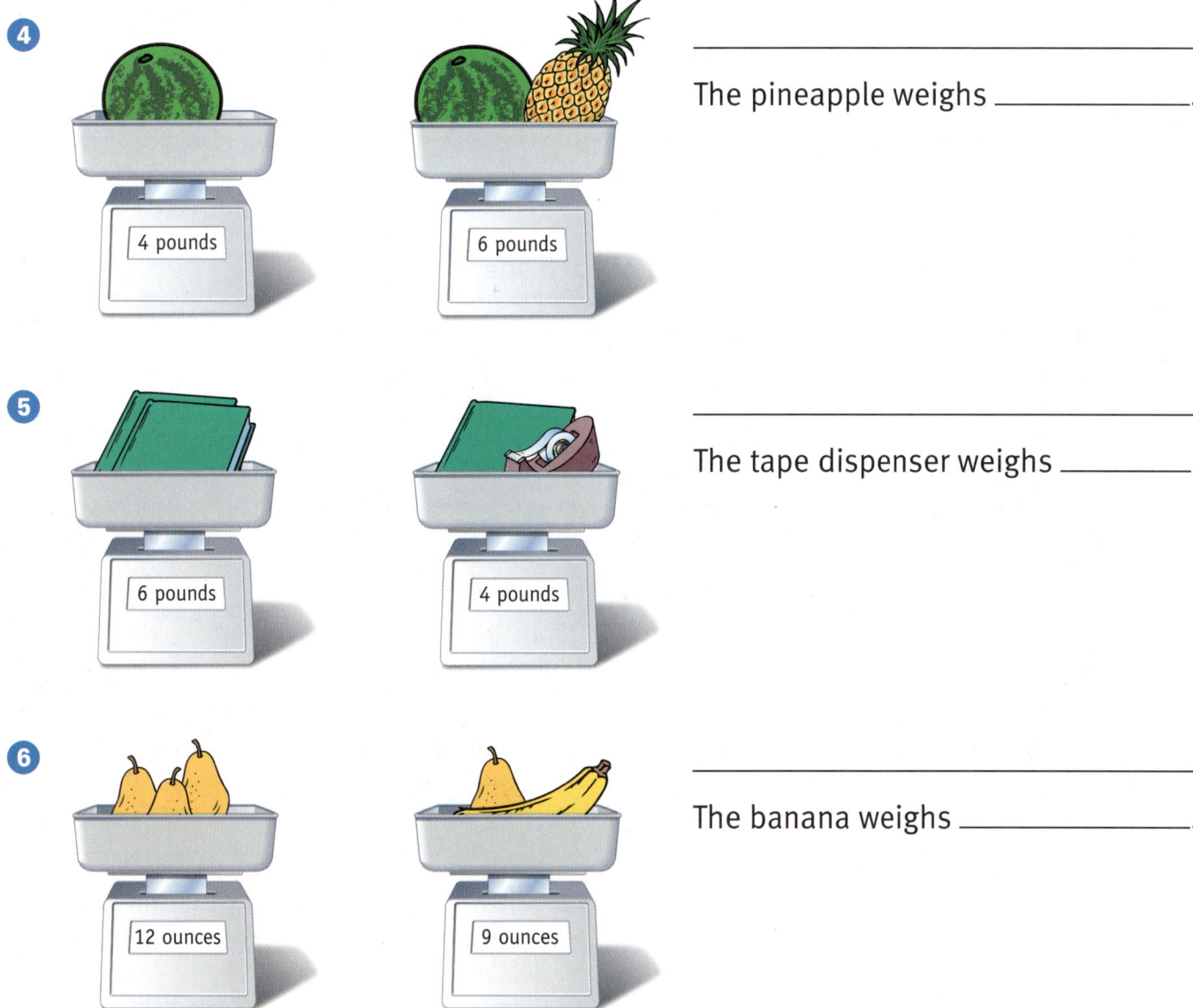

4 The pineapple weighs _____.

5 The tape dispenser weighs _____.

6 The banana weighs _____.

Reflect

How did you decide how much one pear weighs in Problem 6? Explain.

_____
_____
_____

Variables and Equality • Lesson 2

Week 4 — Variables and Equality

Lesson 3

Key Idea
Use reasoning to solve more challenging problems involving weights.

Try This
Answer each question to find the weight of each shape.

1. How much does the pyramid weigh?
   _____

2. How much do the pyramid and cylinder weigh altogether?
   _____

3. How much does the cylinder weigh?
   _____

4. How much do the cylinder and cube weigh altogether?
   _____

5. How much does the cube weigh?
   _____

Practice
Find each unknown weight.

6. The shoe weighs _____.
7. The basket weighs _____.
8. The plant weighs _____.

9. Each tennis ball weighs _____.
10. Each baseball weighs _____.
11. The volleyball weighs _____.

Reflect
Would you be able to find the weight of the volleyball in Problem 7 if you had only the first and third scales? Explain your answer.

_____

Variables and Equality • Lesson 3

Week 4 — Variables and Equality

Lesson 4

> **Key Idea**
> Balance scales can be used to help solve equations.
> When a scale is balanced, both sides are equal.

Try This
Use each balance scale to find two equal weights.

1

The weight of 1 orange is the same as the weight of _____.

2

The weight of 1 toy puppy is the same as the weight of

_____.

3

The weight of 1 box of crayons is the same as the weight of _____

_____.

4

The weight of 4 baseballs is the same as the weight of _____

_____.

Practice
Find each unknown weight. Draw your answer on the scale with the question mark.

5

6

7

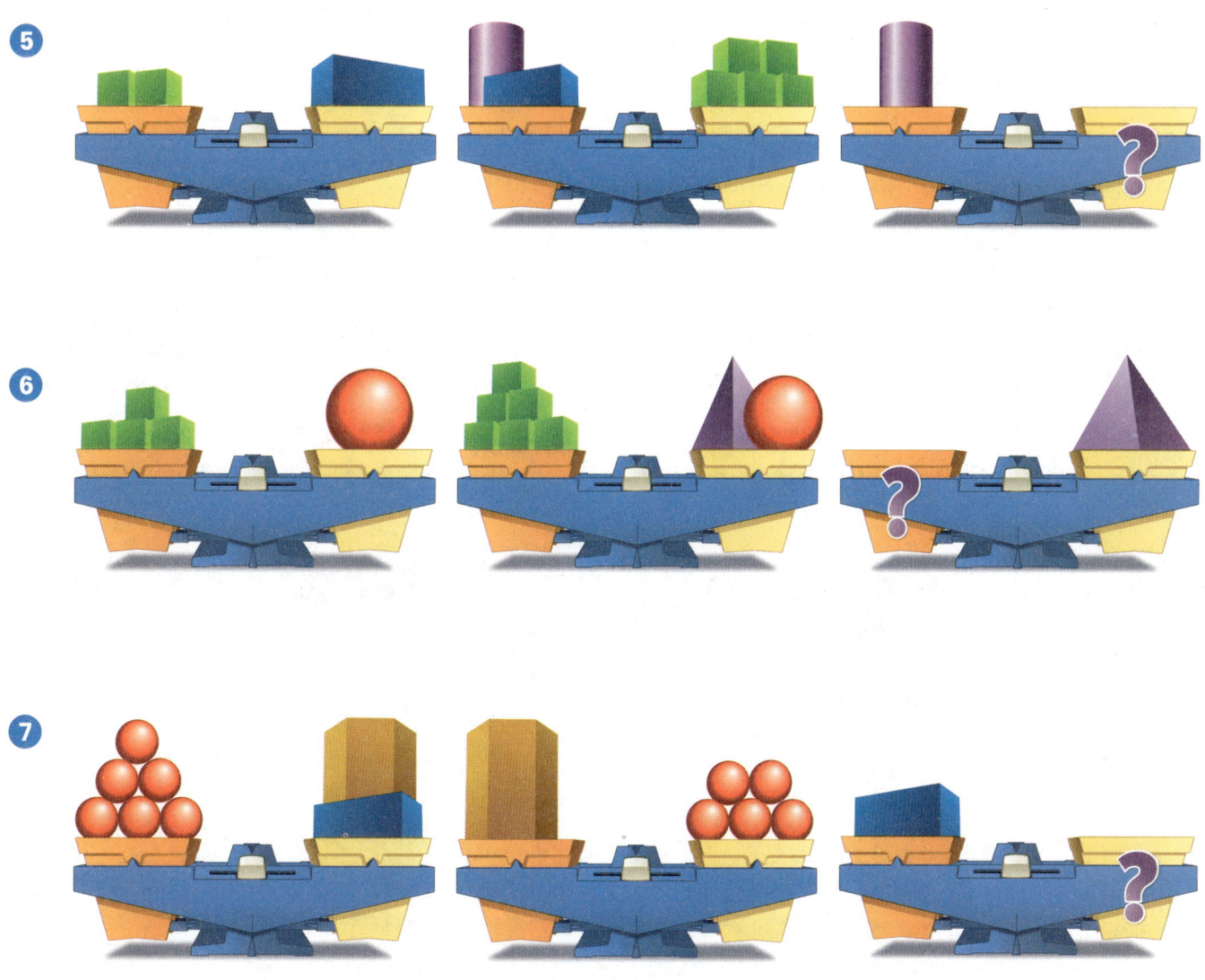

Reflect
What part of a number sentence is represented by the balance scale? Explain.

_____

_____

_____

Variables and Equality • Lesson 4 39

Variables and Equality

Lesson 5 Review

This week you explored equality and unknown values in number sentences. You used shapes to represent missing numbers in an equation. You also related number sentences to weights and balance scales.

Lesson 1 Find the unknown value in each equation.

① $c + 2 = 8$

$c =$ _____

② ☐ + ☐ + ☐ = 9

What is ☐ ?

Lesson 2 Find each unknown weight. Write a number sentence to show your work.

③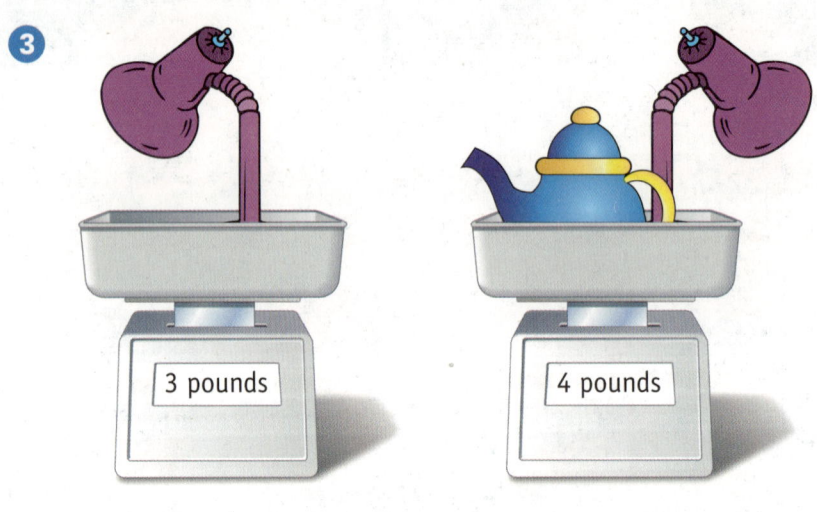

3 pounds   4 pounds

_____

The teapot weighs _____.

Reflect

Explain how to find the values of the unknowns in the number sentence.

☐ + ☐ + ☐ = 21

_____

_____

Lesson 3 ❹

The knife weighs _____.

Lesson 4 Find the unknown weight. Draw your answer on the scale with the question mark.

❺

Reflect

Use shapes from Problem 5 to balance each scale. Draw your answer on the scale with the question mark.

Week 1 Exploring Patterns

Practice

1 Draw the next set in the pattern.

Set 1   Set 2   Set 3   _____

2 Draw the missing set in the pattern.

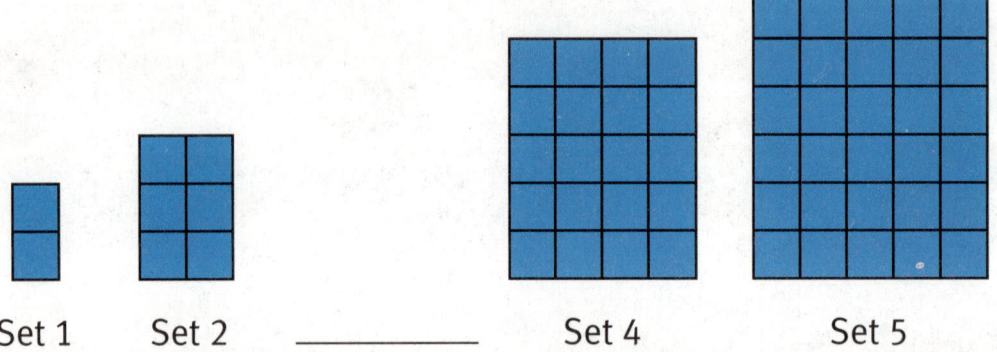

Set 1   Set 2   _____   Set 4   Set 5

3 Tell whether each pattern above is a same-step or a changing-step growing pattern.

_____

_____

4 Design a same-step or changing-step pattern. Show sets 1, 2, and 3, and describe the pattern.

Week 2 Patterns and Relationships

Practice

Complete the table for the pattern shown below.

 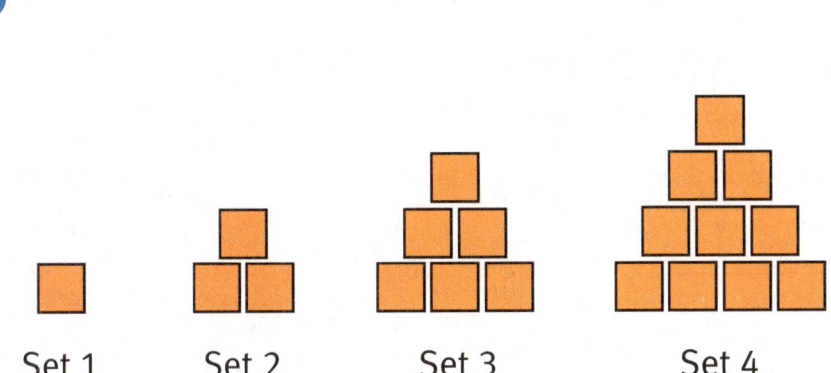

Set	Number of Squares

Set 1   Set 2   Set 3   Set 4

Complete each input/output table.

② Candy bars cost 50¢ each.

Input (number of candy bars)	Output (total cost)

③ Jean can ride her bike 20 miles per hour.

Input (number of hours)	Output (number of miles)

Number Patterns and Relationships • Week 2 Practice

Week 3 Patterns and Graphs

Practice

1 The Booster Club sells gourmet cookies for $1.50 each. Complete the input/output table, and graph the pattern.

Input (number of cookies)	Output (total cost)

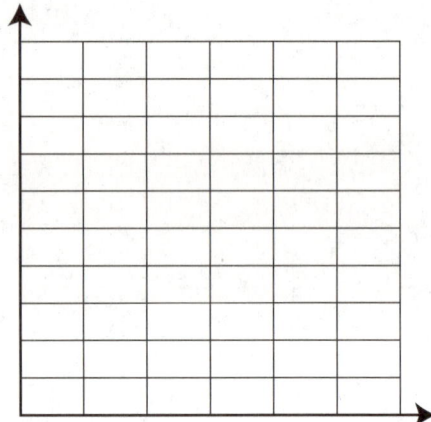

2 Describe the pattern that is shown in the graph.

_____
_____
_____

3 How much would it cost to purchase 10 cookies?

_____

4 What remains the same in the graph?

_____

44 Number Patterns and Relationships • Week 3 Practice

Week 4 — Variables and Equality

Practice

1. $x - 5 = 2$
   $x = $ _____

2. $10 - y = 6$
   $y = $ _____

3. $z + 5 = 15$
   $z = $ _____

4. ☐ + ☐ + ☐ = 9
   What is ☐?

5. $22 - \triangle = \triangle$
   What is △?

6. ◯ + ◯ = 12
   What is ◯?

7. $\triangle + \triangle = 18$
   What is △?

8. $m - 6 = 2$
   $m = $ _____

9. Explain how to find the values of the unknowns in the number sentence
   ☐ + ☐ + ☐ = 27

   _____
   _____
   _____
   _____

Number Patterns and Relationships • Week 4 Practice **45**

Unit 2 Workbook

SRAonline.com

Level F   R53180.01

Unit 2 Workbook
Level F

NUMBER WORLDS

Number Patterns and Relationships

featuring Building Blocks Software

Author
Sharon Griffin
Associate Professor of Education and
Adjunct Associate Professor of Psychology
Clark University
Worcester, Massachusetts

Building Blocks Authors

Douglas H. Clements
Professor of Early Childhood
and Mathematics Education
University at Buffalo
State University of New York, New York

Julie Sarama
Associate Professor of Mathematics Education
University at Buffalo
State University of New York, New York

Contributing Writers
Sherry Booth, *Math Curriculum Developer,* Raleigh, North Carolina
Elizabeth Jimenez, *English Language Learner Consultant,* Pomona, California

Program Reviewers

Jean Delwiche
Almaden Country School
San Jose, California

Cheryl Glorioso
Santa Ana Unified School District
Santa Ana, California

Sharon LaPoint
School District of Indian River County
Vero Beach, Florida

Leigh Lidrbauch
Pasadena Independent School District
Pasadena, Texas

Dave Maresh
Morongo Unified School District
Yucca Valley, California

Mary Mayberry
Mon Valley Education Consortium, AIU 3
Clairton, Pennsylvania

Lauren Parente
Mountain Lakes School District
Mountain Lakes, New Jersey

Juan Regalado
Houston Independent School District
Houston, Texas

M. Kate Thiry
Dublin City School District
Dublin, Ohio

Susan C. Vohrer
Baltimore County Public Schools
Baltimore, Maryland

SRAonline.com

Copyright © 2007 SRA/McGraw-Hill.

All rights reserved. Except as permitted under the United States Copyright Act, no part of this publication may be reproduced or distributed in any form or by any means, or stored in a database or retrieval system, without the prior written permission of the publisher, unless otherwise indicated.

Printed in the United States of America.

Send all inquiries to:
SRA/McGraw-Hill
4400 Easton Commons
Columbus, OH 43219

R53180.01

7 8 9 QPE 12 11 10 09

Photo Credits
2–14 ©PhotoDisc/Getty Images, Inc.;
15 ©Stockbyte/Stockbyte; **16–23** ©PhotoDisc/Getty Images, Inc.; **24** ©Eyewire/Getty Images, Inc.; **25** ©Matt Meadows; **33** ©Stockbyte/Stockbyte

The McGraw·Hill Companies

Contents

Number Patterns and Relationships

Week 1 Exploring Patterns 2

Week 2 Patterns and Relationships 12

Week 3 Patterns and Graphs 22

Week 4 Variables and Equality 32

Week 1 Practice 42

Week 2 Practice 43

Week 3 Practice 44

Week 4 Practice 45

Week 1 — Exploring Patterns

Lesson 1

Key Idea

Same-step patterns are patterns in which the amount of change or growth is the same from one set to the next.

Ask the following questions to help you look for a pattern.

- What is changing from one set to the next?
- What stays the same from one set to the next?
- How would you create the next set?

Try This

Describe what changes in each pattern. Sketch and label the next figure in the pattern.

1

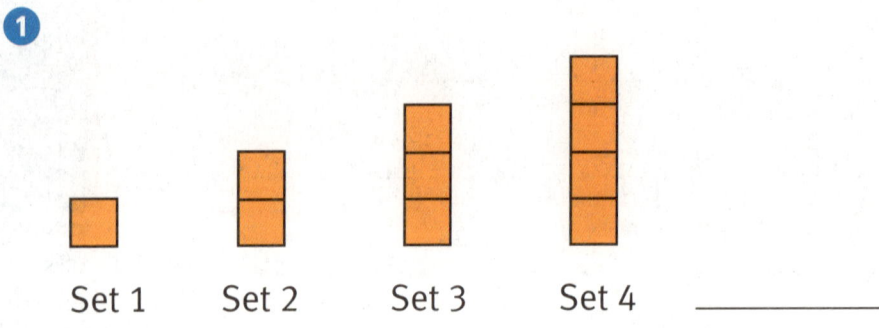

Set 1   Set 2   Set 3   Set 4   _____

_____

_____

2

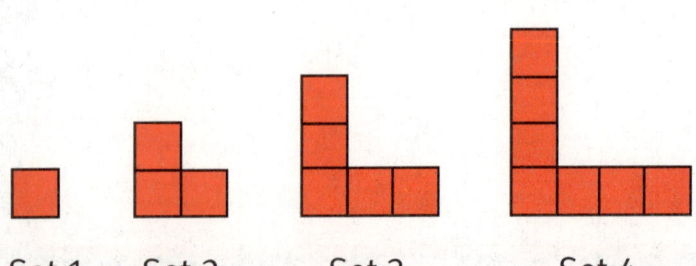

Set 1   Set 2   Set 3   Set 4   _____

_____

_____

2   Number Patterns and Relationships • Week 1

Practice
Sketch and label the next two terms in each pattern.

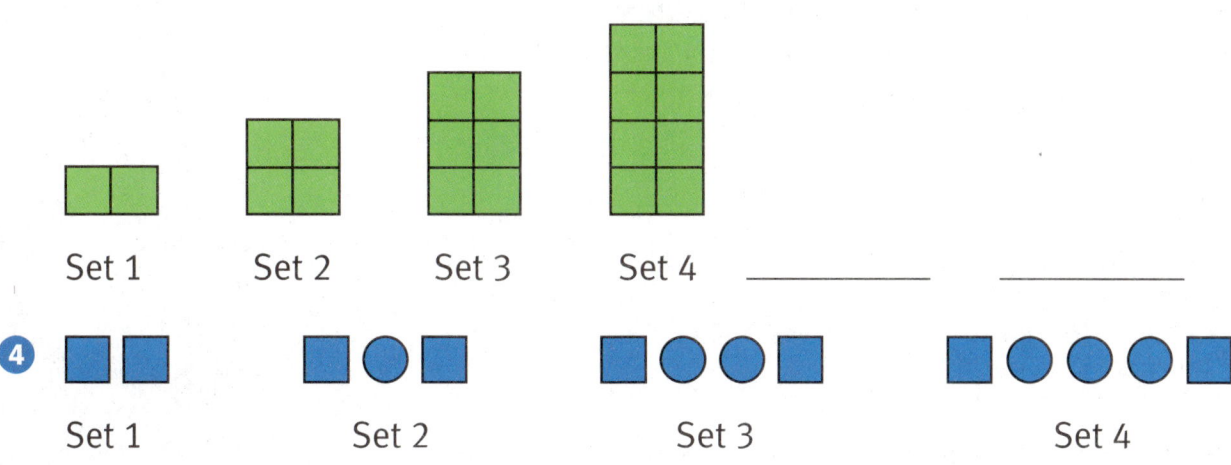

Reflect
Mandy had 5 pennies in her piggy bank on Monday. Each day she put 10 more pennies into the bank. How much money will she have in the piggy bank on Friday? Explain.

Monday	Tuesday	Wednesday	Thursday	Friday
5¢	15¢	25¢	35¢	_____

_____
_____
_____
_____

Week 1

Exploring Patterns

Lesson 2

Key Idea

Changing-step patterns are patterns in which the amount of change or growth changes in a regular and predictable way from one set to the next.

Ask the following questions to help you identify a changing-step pattern:

- What changes from one set to the next?
- What stays the same from one set to the next?
- How would you create the next set?

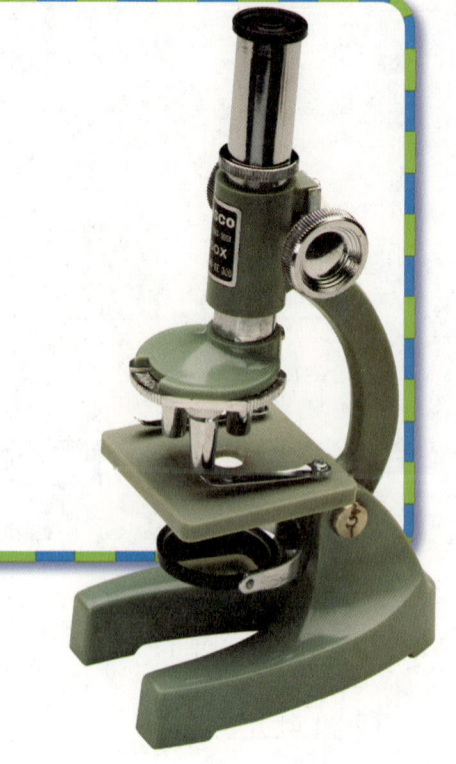

Try This

Describe the changes in each pattern. Then sketch and label the next set in the pattern.

1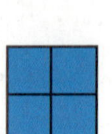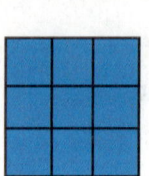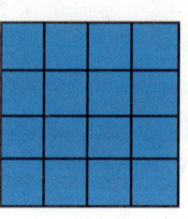

Set 1   Set 2   Set 3   Set 4   _____

_____

_____

2

Set 1   Set 2   Set 3   _____

_____

_____

4 Number Patterns and Relationships • Week 1

Practice
Sketch and label the next term in the pattern.

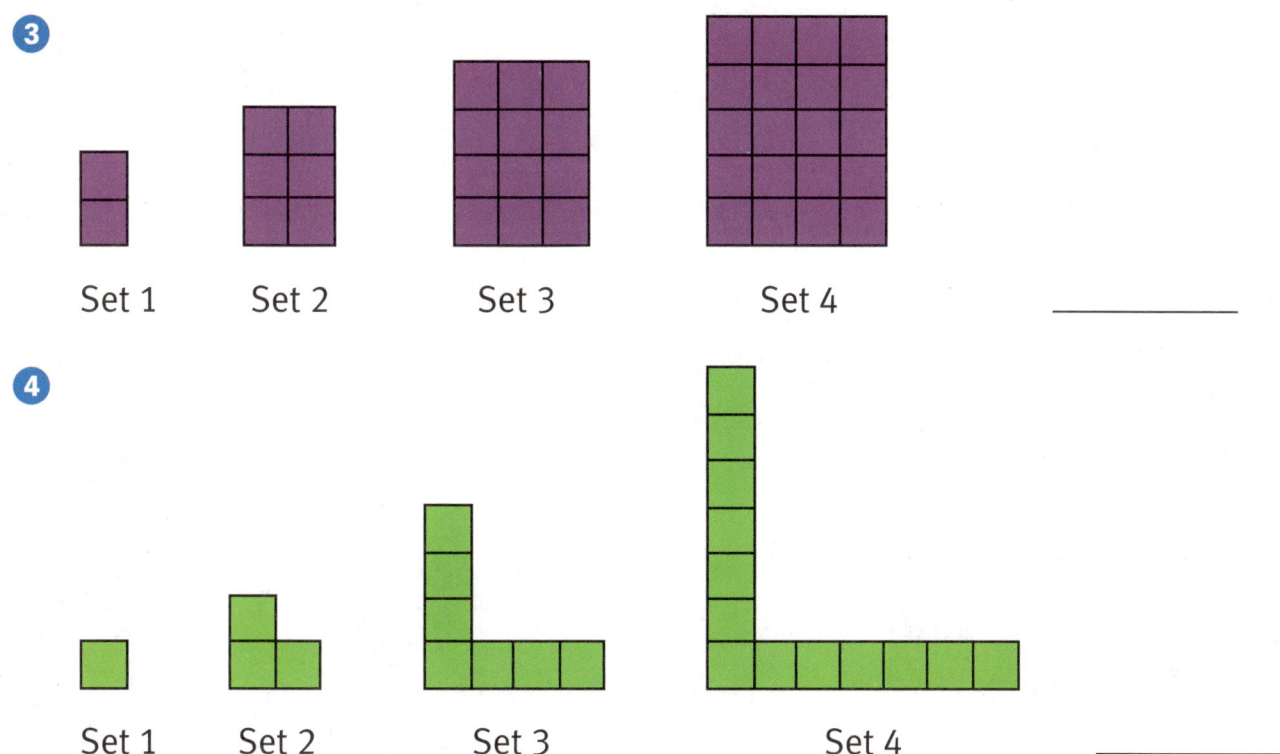

3 Set 1, Set 2, Set 3, Set 4, _____

4 Set 1, Set 2, Set 3, Set 4, _____

Reflect
Look for a pattern in the table. If the pattern continues, how many chin-ups will Jarrod do on Friday? Explain your answer.

Jarrod's Chin-up Chart				
Monday	Tuesday	Wednesday	Thursday	Friday
8 chin-ups	10 chin-ups	13 chin-ups	17 chin-ups	

_____
_____
_____

Exploring Patterns • Lesson 2

Week 1 Exploring Patterns

Lesson 3

Key Idea
You can use clues from the sets given in a pattern to identify missing sets.

Try This
Look for the changes in each pattern. Then draw and label the missing set.

① Set 1   Set 2   _____   Set 4   Set 5

② Set 1   Set 2   _____   Set 4

③ Set 1   _____   Set 3   Set 4   Set 5

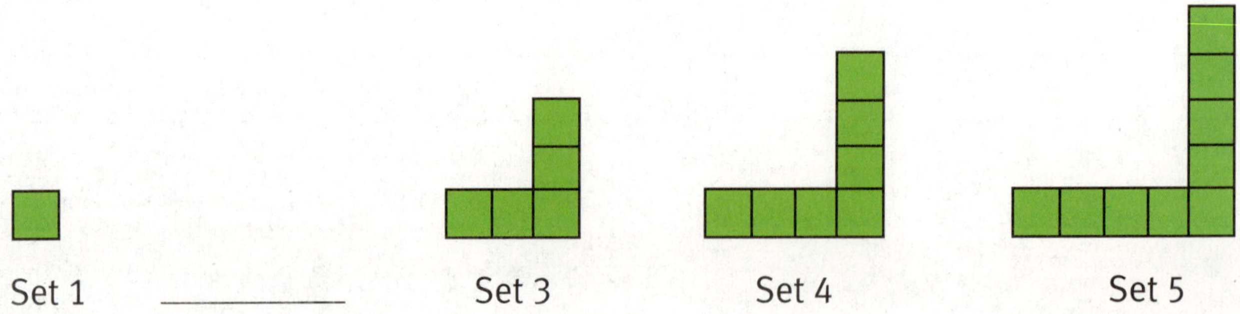

Practice

Look for the changes in each pattern. Tell whether it is a same-step growing pattern or a changing-step growing pattern. Then draw and label the missing set.

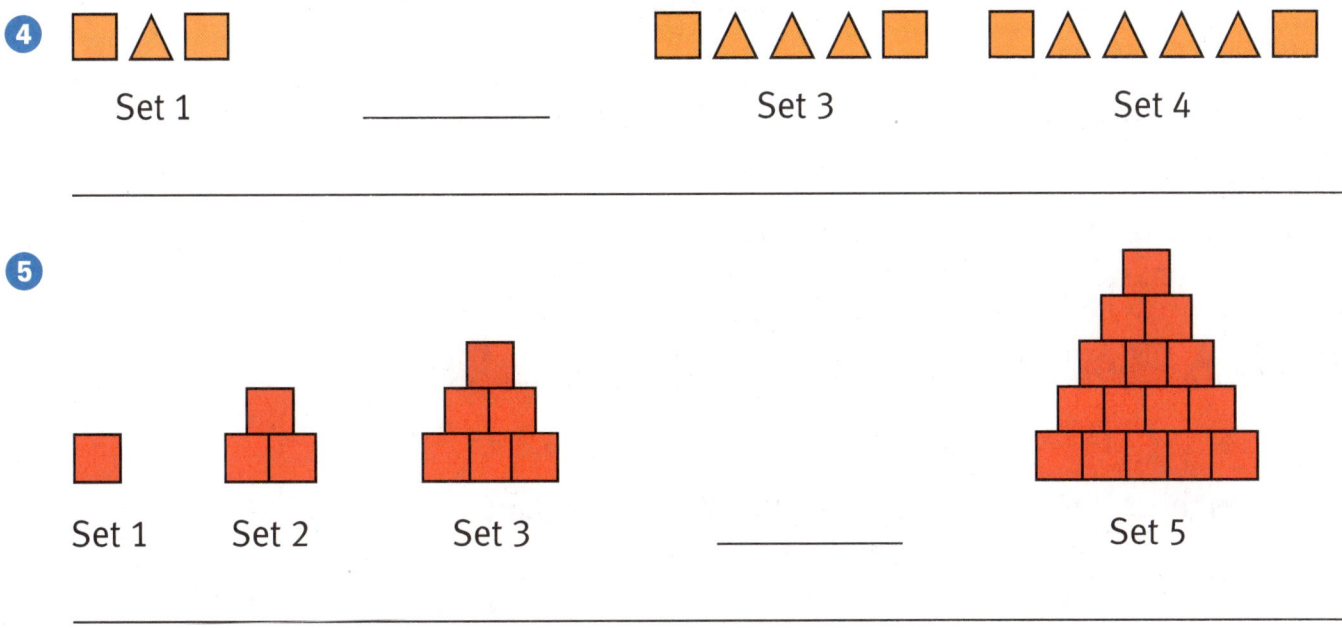

④ ▢▵▢   ▢▵▵▵▢   ▢▵▵▵▵▢
   Set 1   _____   Set 3   Set 4

⑤ Set 1   Set 2   Set 3   _____   Set 5

Reflect

Do you think it is easier to find a missing set in a same-step or a changing-step growing pattern? Explain.

_____
_____
_____
_____

Exploring Patterns • Lesson 3   **7**

Week 1 — Exploring Patterns

Lesson 4

> **Key Idea**
> You can use pattern blocks to create your own growing patterns.

Try This
Draw and label the next set for each pattern. Use the pattern to answer the questions.

1

   Set 1   Set 2   Set 3   _____

   a. Is this a same-step or a changing-step growing pattern?

   _____

   b. How is the pattern changing from set to set?

   _____

2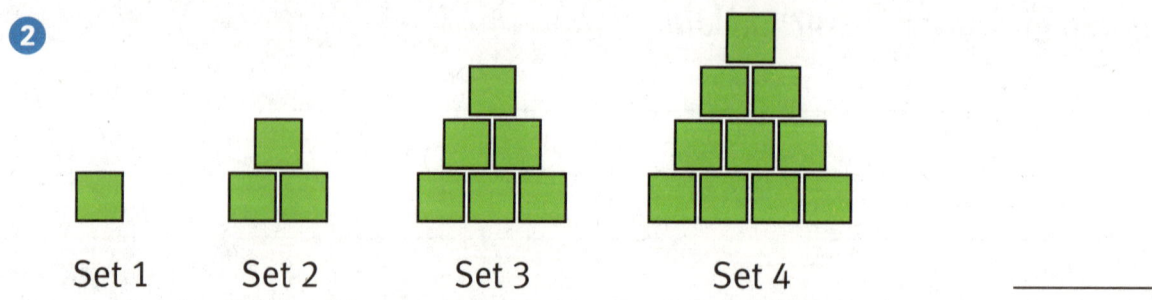

   Set 1   Set 2   Set 3   Set 4   _____

   a. Is this a same-step or a changing-step growing pattern?

   _____

   b. How is the pattern changing from set to set?

   _____

   _____

8 Number Patterns and Relationships • Week 1

Practice

Draw your own pattern in the space below. Exchange your pattern with a partner, and have your partner answer each question.

3 Is this a same-step or a changing-step growing pattern?

_____

4 How is the pattern changing from set to set?

_____

5 What would the next set of the pattern look like? Describe it.

_____

Reflect

Use different shapes to create the first few sets of a pattern. Explain how the pattern changes from set to set.

_____

_____

Exploring Patterns • Lesson 4

Week 1 — Exploring Patterns

Lesson 5 Review

This week you explored patterns. You examined same-step patterns and changing-step patterns.

Lesson 1 Sketch and label the next set in the pattern.

❶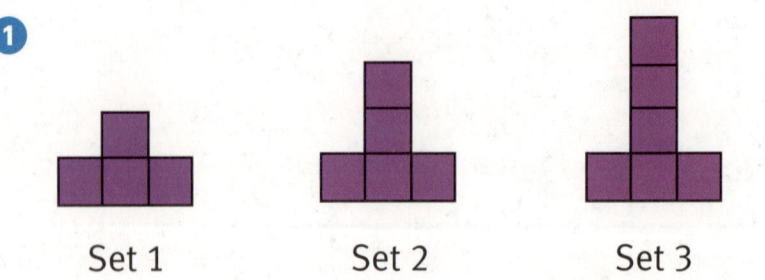

❷

Lesson 2 Sketch and label the next set in the pattern.

❸

Reflect
Tell whether each pattern above is a same-step or a changing-step growing pattern.

_____
_____
_____

10 Number Patterns and Relationships • Week 1

Lesson 3 Sketch and label the missing set in the pattern.

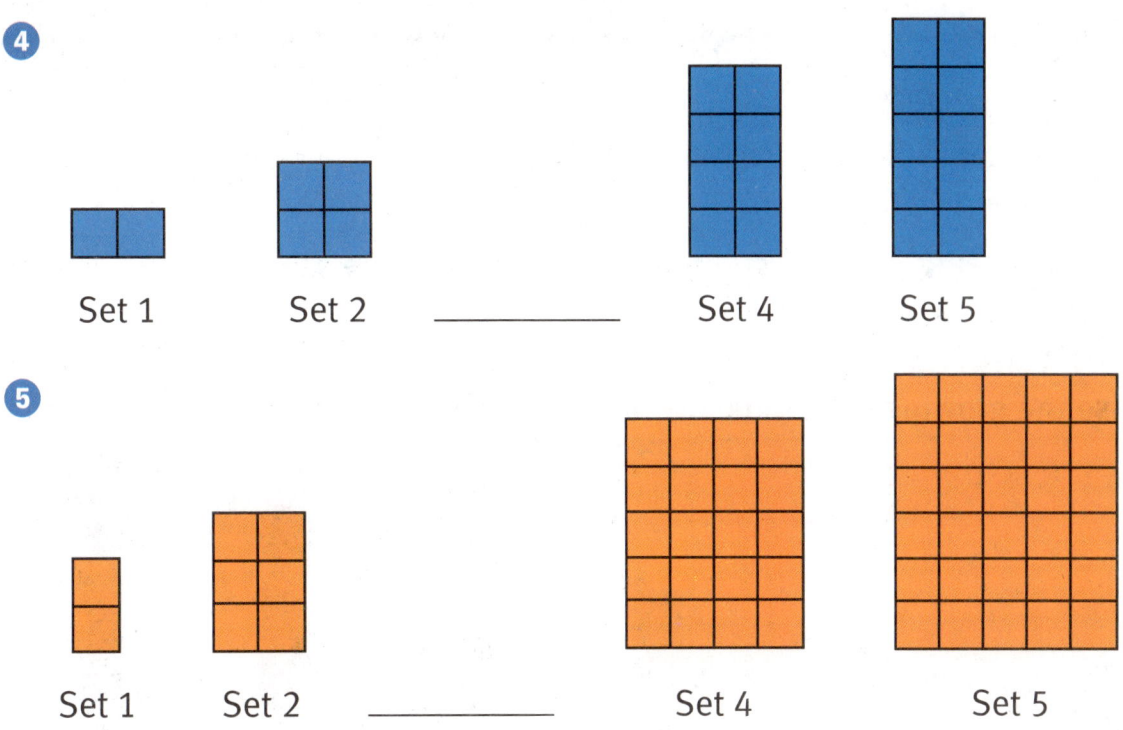

④ Set 1 Set 2 _____ Set 4 Set 5

⑤ Set 1 Set 2 _____ Set 4 Set 5

Lesson 4 ⑥ Use pattern blocks to design Set 1 of a pattern. Then show Sets 2 and 3.

Reflect
Describe how to find the missing set. Then draw and label the missing set.

Set 1   Set 2   _____   Set 4

_____

_____

_____

Exploring Patterns • Lesson 5 Review

Week 2 Patterns and Relationships

Lesson 1

Key Idea
You can use numbers to represent the growth in patterns.

Try This
Use the growing pattern below to answer each question.

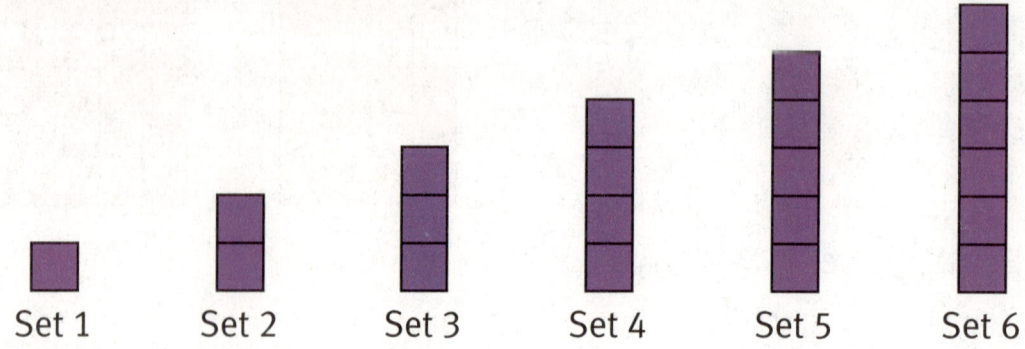

1. Use the pattern to complete the table below.

Set	Number of Squares
1	
2	
3	
4	
5	
6	

2. What patterns do you notice in the table between the set number and the number of squares?

_____

_____

3. Use the pattern from Problem 2 to predict the number of squares in the tenth set without building or drawing all the sets up to the tenth.

_____

12 Number Patterns and Relationships • Week 2

Practice
Use the pattern to complete the table.

④

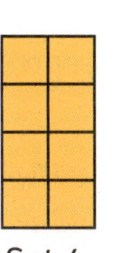
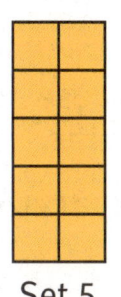

Set 1   Set 2   Set 3   Set 4   Set 5

Set	Number of Squares
1	
2	
3	
4	
5	

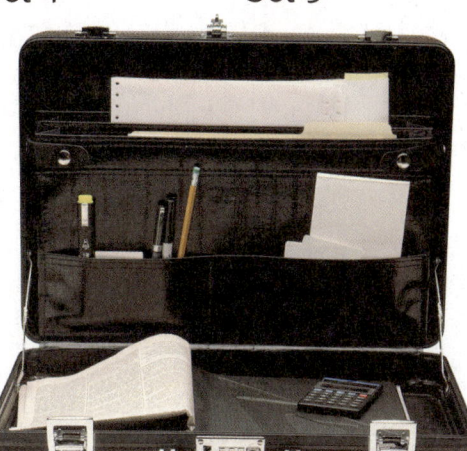

⑤ What patterns do you notice in the table above between the set number and the number of squares?

_____

⑥ Use the pattern from Problem 5 to predict the number of squares in the tenth set without building or drawing all of the sets up to the tenth.

_____

Reflect
When you use a pattern table to help you create a pattern with shapes, what is the smallest number of sets you need to figure out the pattern? Explain.

_____

_____

_____

Patterns and Relationships • Lesson 1

Week 2 — Patterns and Relationships

Lesson 2

Key Idea
You can use numbers to represent the growth in patterns.

Try This
Use the growing pattern below to answer each question.

Set 1   Set 2   Set 3   Set 4   Set 5

1. Is this a same-step or a changing-step growth pattern?
   _____

2. Describe the growth shown in the pattern.
   _____

3. How many squares were used to create each set?
   _____

4. Use your answers to complete the table below.

Set	Number of Squares
1	
2	
3	
4	
5	

14 Number Patterns and Relationships • Week 2

Practice
Use the pattern to complete the table.

5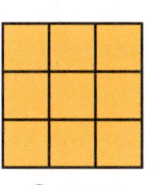

Set 1   Set 2   Set 3   Set 4   Set 5

Set	Number of Squares
1	
2	
3	
4	
5	

6 What patterns do you notice in the table above between the set number and the number of squares?

_____

7 Use the pattern to predict the number of squares in the tenth set without building or drawing all of the sets up to the tenth.

_____

Reflect
What do you notice about the table for a changing-step pattern? How does it compare with the table for a same-step pattern?

_____

_____

_____

Patterns and Relationships • Lesson 2

Week 2 — Patterns and Relationships

Lesson 3

Key Idea

The pattern tables are examples of input/output tables.

For each input value (the set number), there is a certain output value (number of squares).

Input/output tables can also be used to model real-world situations.

Try This

Thomas mows lawns in his neighborhood to earn money. He earns $8 for each lawn he mows. Use the input/output table to answer each question.

Input (lawns mowed)	Output (money earned)
1	$8
2	$16
3	$24
4	$32
5	$40

① How much money would Thomas earn if he mowed 6 lawns? Explain how you found your answer.

_____

_____

_____

② Write a mathematical rule for determining the amount of money Thomas will earn for any given number of lawns mowed.

_____

Practice
Complete the table and answer each question.

3 Movie tickets cost $6 each. Complete the input/output table.

Input (number of tickets)	Output (total cost)
1	$6
2	
3	
4	
5	

4 How much would it cost to buy 4 movie tickets?

_____

5 Write a mathematical rule for determining the total cost of tickets for any given number of tickets.

_____

_____

6 For every hour Lisa drives, she uses 2 gallons of gasoline. Her gas tank holds 18 gallons when it is full. Complete the input/output table.

Input (hours of driving)	Output (gas remaining in her tank)
1	16 gallons
2	14 gallons
3	
4	
5	

7 How much gasoline is in Lisa's tank after 5 hours of driving?

_____

8 Write a mathematical rule for determining the amount of gas remaining for any given number of hours driven.

_____

_____

Reflect
What is different about the lawn mowing input/output table and the gasoline input/output table?

_____

_____

_____

Patterns and Relationships • Lesson 3   **17**

Week 2 — Patterns and Relationships

Lesson 4

Key Idea
You can use input/output tables to help you make choices.

Try This

Anna's neighbors have hired her to pet sit their dog for seven days. They have offered two different options for being paid.

- **Option 1:** Anna receives $10 for the first day and an additional $2 per day after the first day.

- **Option 2:** Anna receives $1 for the first day. Every day after the first day she receives an additional amount that is $1 more than the previous day.

1. Complete the input/output tables for each option.

Option 1	
Day	Total Amount Earned
1	$10
2	$12
3	
4	
5	
6	
7	

Option 2	
Day	Total Amount Earned
1	$1
2	$3
3	$6
4	
5	
6	
7	

2. If Anna chooses Option 1, how much money will she be paid? If Anna chooses Option 2, how much money will she be paid?

Practice

Jim was hired to do yard work for his neighbor. The neighbor expects the work to last 5 days, but it could last 7 days. Payment options are as follows:

- **Option 1:** Jim receives $12 for the first day and $2 per day after the first day.

- **Option 2:** Jim receives $2 for the first two days. Every day thereafter he receives an amount that is $1 more than the previous day.

Complete the tables to find Jim's total earnings for the week for each option.

❸ Complete the table for each option.

Option 1		
Day	Amount Earned for the Day	Total Earnings
1	$12	
2		
3		
4		
5		
6		
7		

Option 2		
Day	Amount Earned for the Day	Total Earnings
1	$2	
2		
3		
4		
5		
6		
7		

❹ How much will Jim earn for 7 days if he chooses Option 1? _____

❺ How much will Jim earn for 7 days if he chooses Option 2? _____

❻ If the job is for only 5 days, which option will pay better? _____

Reflect

What kind of growth pattern is shown by Option 1? What kind of growth pattern is represented by Option 2?

Week 2 — Patterns and Relationships

Lesson 5 Review

This week you explored patterns and relationships. You looked at how visual patterns can be related to number patterns. You also learned about input/output tables and solving problems.

Lessons 1 and 2

Complete the table for the pattern shown below.

1

Set 1   Set 2   Set 3   Set 4

Set	Number of Cubes
1	
2	
3	
4	

Reflect
How is the pattern changing?
_____

20 Number Patterns and Relationships • Week 2

Lesson 3 Complete the input/output table.

2 The bookstore sells pencils for 15¢ each.

Input (number of pencils)	Output (total cost)
1	15¢
2	
3	
4	

Lesson 4 A bathtub holds 60 gallons of water. When the drain plug is pulled, 12 gallons drain from the tub each minute.

3 How long does it take for the tub to fully drain?

_____

Input (number of minutes)	Output (water remaining in the tub)
0	60 gallons
1	
2	
3	
4	
5	

Reflect
How many gallons of water are left in the tub 1 minute after the plug is pulled? How many gallons of water are left in the tub 3 minutes after the plug is pulled? Is this an example of same-step pattern or a changing-step pattern?

_____
_____
_____

Week 3 — **Patterns and Graphs**

Lesson 1

> **Key Idea**
> Patterns can be represented with pictures, rules, and tables. They can also be represented with graphs.

Try This
Below is a graph that shows how far Lisa can drive, depending on the number of gallons of gasoline in the car's tank.

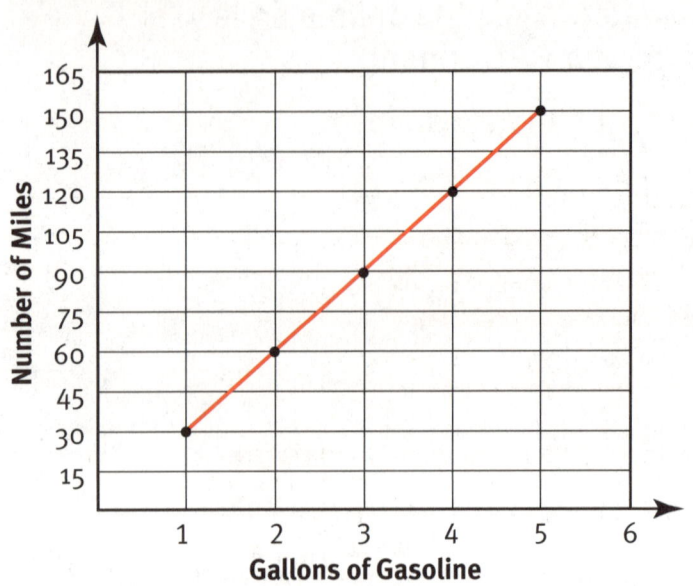

① Which axis represents the number of miles Lisa can drive?
_____

② Which axis represents the amount of gasoline in Lisa's car?
_____

③ As the number of gallons of gasoline increases, what happens to the distance that Lisa can drive? Is this increase a same-step increasing pattern or a changing-step pattern?
_____
_____

④ Describe the pattern shown in the graph.
_____
_____
_____

Practice

The graph shows the money Thomas made mowing lawns.

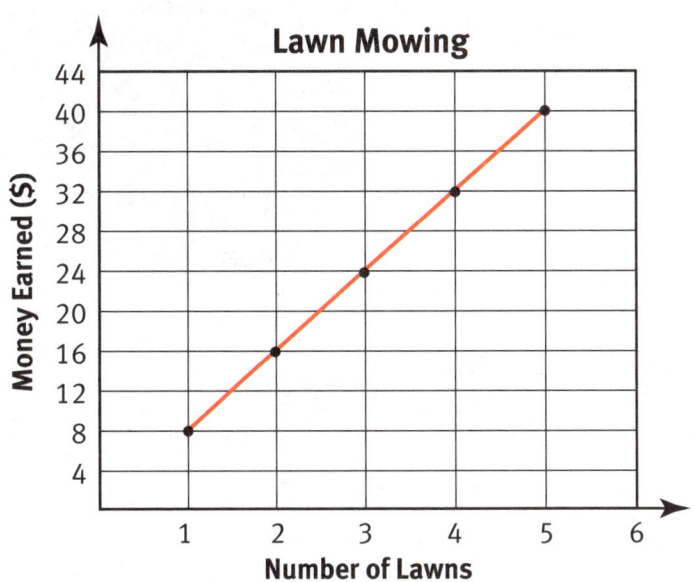

5 The data points are connected to help you see the trend of the data. As the number of lawns mowed increases, what happens to the amount of money earned?

_____

6 By connecting the data points, you are showing that the data is continuous. Should these data points be connected? Explain your answer.

_____

_____

_____

_____

Reflect

Create an input/output table, using the information above.

Patterns and Graphs

Lesson 2

Key Idea
When creating a graph, be sure to label the axes and give it a title.

Try This
Follow the steps to create a graph of the pattern.

Input (movie tickets)	Output (total cost)
1	$6
2	$12
3	$18
4	$24
5	$30

Step 1 Label the horizontal axis and the vertical axis.

Step 2 Plot a point for each pair of numbers in the table.

Step 3 Give your graph a title.

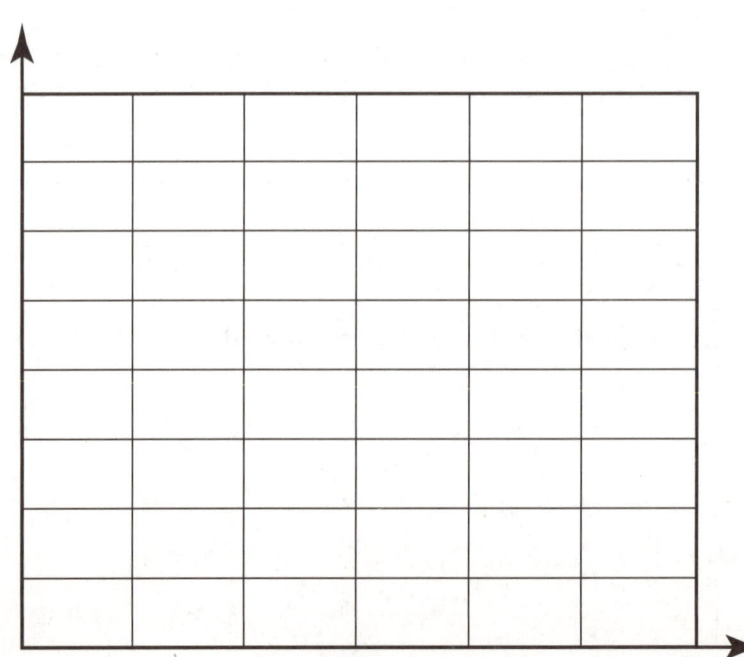

Practice

Mandy is babysitting for her neighbors. Graph the pattern shown in the input/output table.

Input (number of hours)	Output (money earned)
1	$5
2	$10
3	$15
4	$20
5	$25

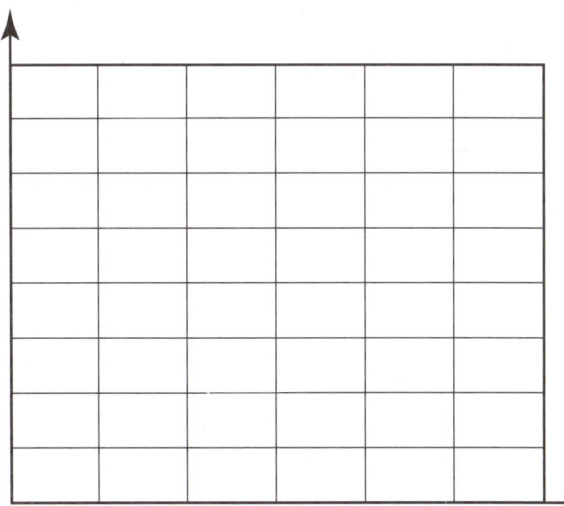

① Describe the pattern shown in the graph.

_____

② How much does Mandy earn for babysitting 6 hours?

_____

Reflect

Can you determine the rule for a pattern by just looking at the graph? Explain and give an example.

_____

_____

_____

Patterns and Graphs • Lesson 2

Week 3 Patterns and Graphs

Lesson 3

> **Key Idea**
> You can use graphs to compare two related patterns.

Try This

Create a graph for each input/output table. Answer each question.

Milk Cartons Sold	
Input (day)	Output (milk sold for the week)
Monday	25 cartons
Tuesday	50 cartons
Wednesday	75 cartons
Thursday	100 cartons
Friday	125 cartons

Milk Cartons Left	
Input (day)	Output (milk cartons left in the cafeteria)
Monday	125 cartons
Tuesday	100 cartons
Wednesday	75 cartons
Thursday	50 cartons
Friday	25 cartons

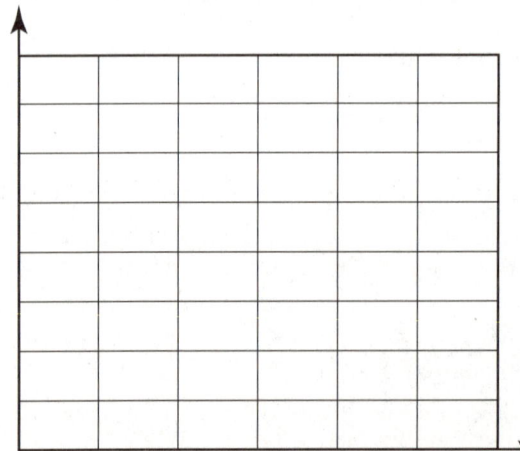
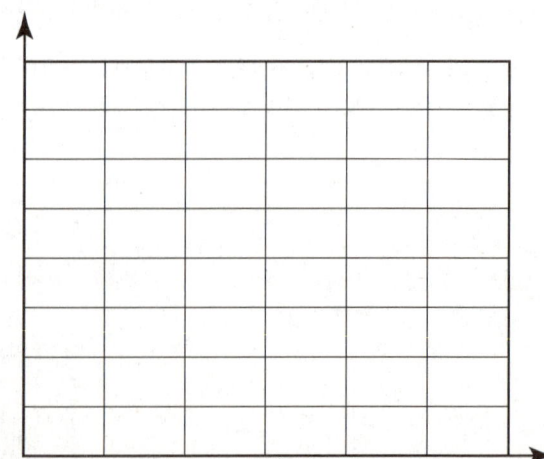

❶ Which arrow line or axis represents the day of the week? Which represents the number of milk cartons left?

_____

_____

26 Number Patterns and Relationships • Week 3

Practice
Use your graphs from Try This to answer each question.

2 Describe the pattern shown in the first graph.
_____

3 Describe the pattern shown in the second graph.
_____

4 Which of the graphs shows a growing pattern?
_____

5 Are these graphs same-step patterns or changing-step patterns?
_____

6 What stays the same in the first graph? What changes?
_____

7 What stays the same in the second graph? What changes?
_____

8 How are the two graphs related?
_____
_____

Reflect
Can you show both patterns on the same graph? Explain.
_____
_____
_____

Patterns and Graphs • Lesson 3

Week 3

Patterns and Graphs

Lesson 4

Key Idea
You can use graphs to tell a story or make an informed decision.

Try This
Choose the story that belongs with each graph.

Story A Melissa rode her bike for 40 minutes. The table shows the distance she traveled.

Time	10 minutes	20 minutes	30 minutes	40 minutes
Distance	3 miles	6 miles	9 miles	12 miles

Story B Mrs. Swanson walked her dog for 40 minutes. The table shows the number of blocks she covered.

Time	10 minutes	20 minutes	30 minutes	40 minutes
Distance	6 blocks	12 blocks	18 blocks	24 blocks

❶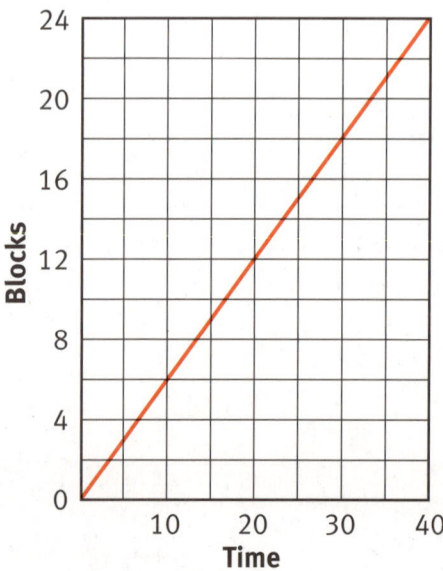

❷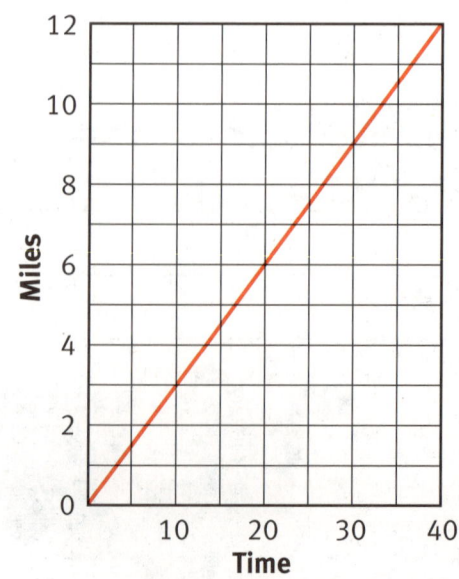

_____ _____

28 Number Patterns and Relationships • Week 3

Practice
Create a graph for the data in the table.

3
Number of Books	1	2	4	6
New Vocabulary Words	2	4	8	12

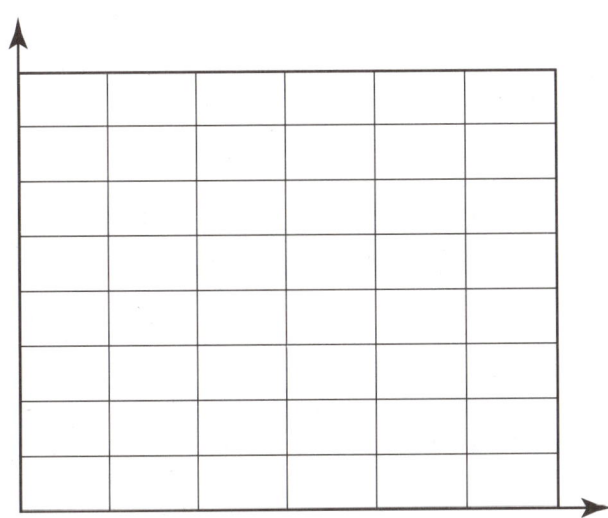

Reflect
Create a table that compares the outside temperature throughout the morning and afternoon of a winter day. Graph the data in the table.

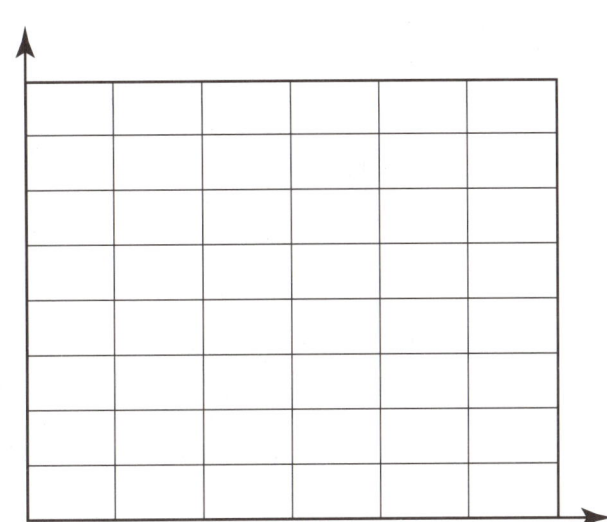

Patterns and Graphs • Lesson 4   29

Week 3 — Patterns and Graphs

Lesson 5 Review

This week you explored how patterns look in graphs. You used input/output tables and stories to create graphs. You also used graphs to answer questions about the pattern and data.

Lessons 1 and 2

The bookstore sells school sweatshirts for $10 each. Graph the pattern shown in the input/output table.

Input (number of sweatshirts)	Output (total cost)
1	$10
2	$20
3	$30
4	$40
5	$50

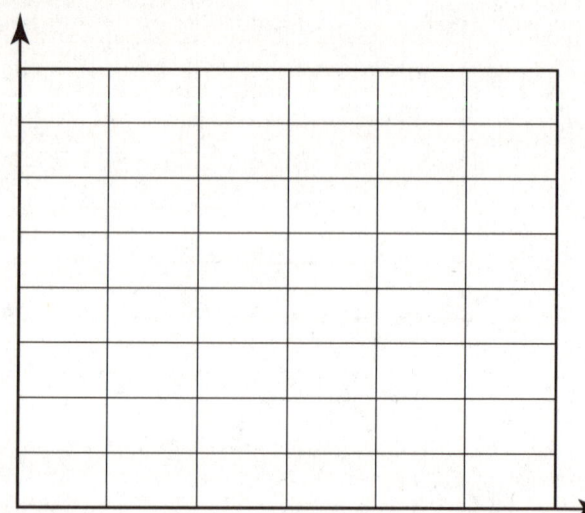

❶ Describe the pattern that is shown in the graph.

_____

_____

❷ How much would it cost to purchase eight sweatshirts?

_____

Reflect
What is staying the same in the graph?

_____

30 Number Patterns and Relationships • Week 3

Lesson 3 Graph the pattern shown in the input/output table.

Input (number of cars)	Output (total passengers)
1	4
2	8
3	12
4	16
5	20

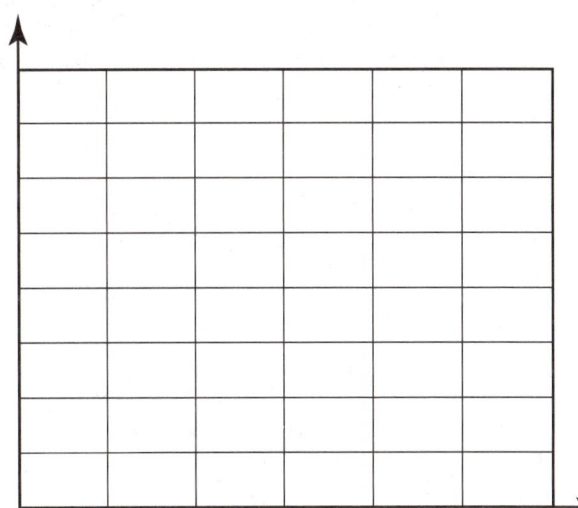

Lesson 4 Use the graph to answer each question.

❸ Which story matches the graph? Circle A or B.

A. A shoe store sold 20 pairs of shoes on Monday. The store sold no shoes on Tuesday because it was closed. On Wednesday and Thursday, 30 pairs of shoes were sold.

B. A hot air balloon rose to 250 feet. It stayed there for a while and then rose to 500 feet. After a little while longer, the balloon began its descent.

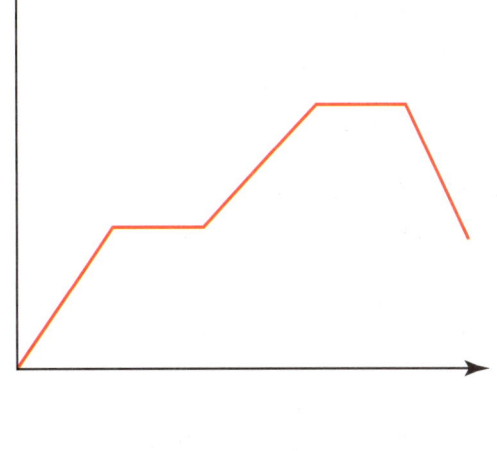

Reflect
What label would you put on the horizontal axis? What label would you put on the vertical axis?

_____

Week 4 — **Variables and Equality**

Lesson 1

Key Idea

An **equation** is a number sentence which states that two mathematical expressions are equal.

$2 + 3 = 5$ $11 - 4 = 7$ $6 - 2 = 3 + 1$

Sometimes equations have unknown values. You can show unknown values with pictures, boxes, or letters.

$4 + \square = 12$ $b - 9 = 5$

Try This

Find the unknown value in each equation. Substitute values into the equation until you have a true number sentence.

1. $\square + 6 = 8$
   What is $\square$?

2. $\triangle + 1 = 7$
   What is $\triangle$?

3. $4 + \triangle = 8$
   What is $\triangle$?

4. $5 - \bigcirc = 2$
   What is $\bigcirc$?

5. $a + 8 = 10$
   $a = \underline{}$

6. $12 - 7 = z$
   $z = \underline{}$

Practice

Find the unknown value in each equation. The same shapes represent the same value.

7. $\square + \square = 10$
   What is $\square$?

8. $\bigcirc + \bigcirc = 2$
   What is $\bigcirc$?

9 △ + △ = 6

What is △ ?

10 ⬡ + ⬡ + ⬡ = 15

What is ⬡ ?

Find the unknown value in each equation.

11 9 − ☐ = 3

What is ☐ ?

12 8 − n = 6

n = _____

13 t − 7 = 2

t = _____

14 14 + ⬡ = 19

What is ⬡ ?

15 x + 4 = 12

x = _____

16 17 − y = 12

y = _____

Reflect

What values of ☐ and △ make a true number sentence? Is there more than one correct answer? Explain.

☐ − △ = 3

_____
_____
_____

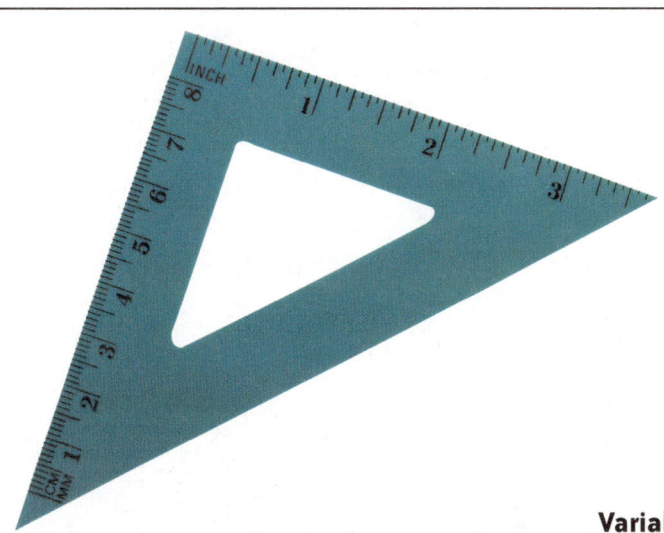

Variables and Equality • Lesson 1   33

Week 4: Variables and Equality

Lesson 2

> **Key Idea**
> You can use the idea of weights to help solve equations.

Try This

Answer each question to find the weight of the toy car.

3 pounds

8 pounds

1. How much does the piggy bank weigh?
   _____

2. How much do the piggy bank and toy car weigh altogether?
   _____

3. Fill in the blanks below to help you find the weight of the toy car.

   Piggy bank = 3

   Piggy bank + toy car = 8

   _____ + toy car = 8

   _____ + _____ = 8

   Toy car = _____

   The toy car weighs _____.

Practice

Find each unknown weight. Write a number sentence to show your work.

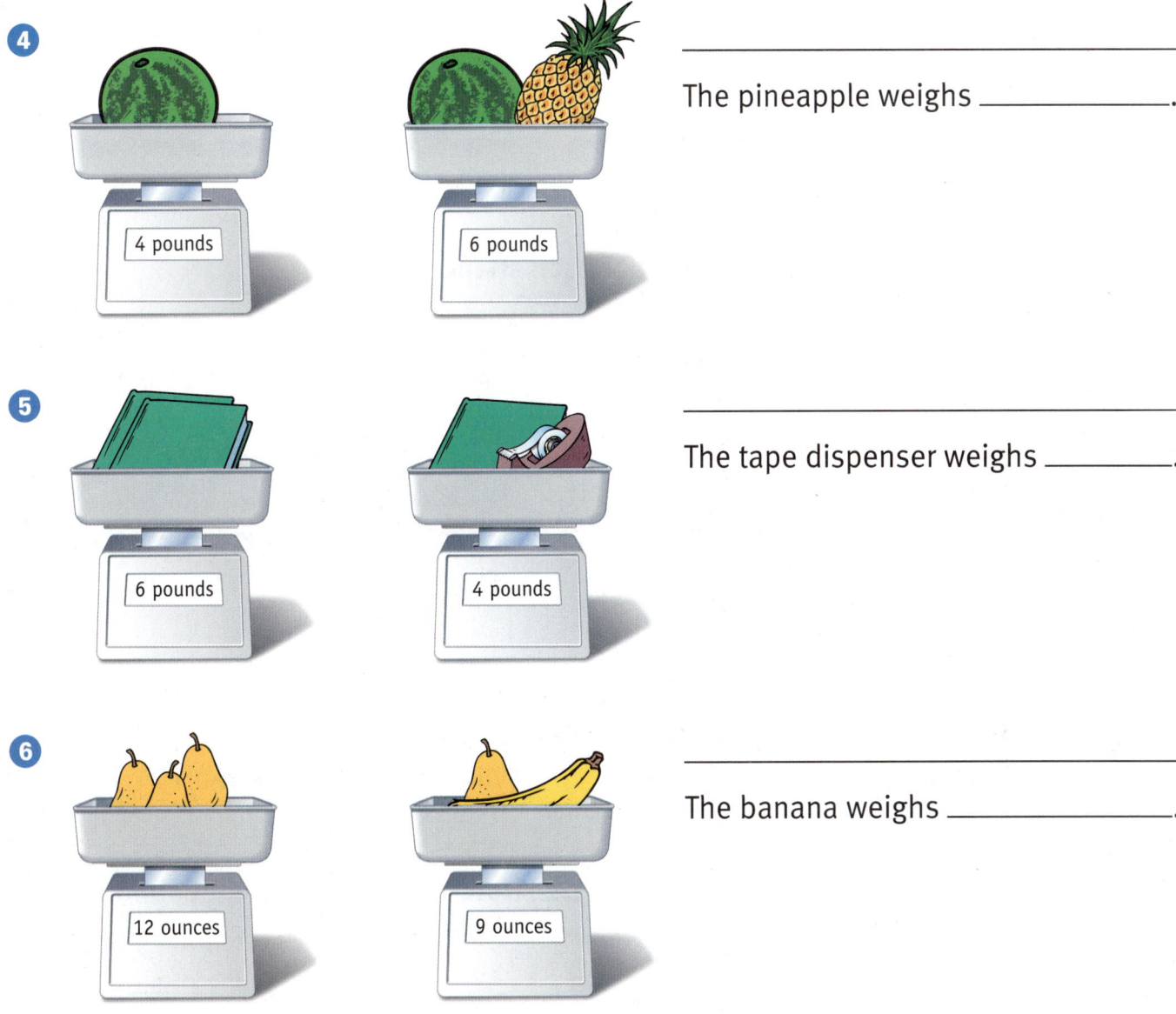

④ The pineapple weighs _____.

⑤ The tape dispenser weighs _____.

⑥ The banana weighs _____.

Reflect

How did you decide how much one pear weighs in Problem 6? Explain.

_____

_____

_____

Variables and Equality • Lesson 2

Week 4 — **Variables and Equality**

Lesson 3

> **Key Idea**
> Use reasoning to solve more challenging problems involving weights.

Try This
Answer each question to find the weight of each shape.

4 pounds   9 pounds   8 pounds

1 How much does the pyramid weigh?
_____

2 How much do the pyramid and cylinder weigh altogether?
_____

3 How much does the cylinder weigh?
_____

4 How much do the cylinder and cube weigh altogether?
_____

5 How much does the cube weigh?
_____

Practice
Find each unknown weight.

6 The shoe weighs _____.

7 The basket weighs _____.

8 The plant weighs _____.

9 Each tennis ball weighs _____.

10 Each baseball weighs _____.

11 The volleyball weighs _____.

Reflect
Would you be able to find the weight of the volleyball in Problem 7 if you had only the first and third scales? Explain your answer.

_____

Variables and Equality • Lesson 3

Week 4 — Variables and Equality

Lesson 4

Key Idea
Balance scales can be used to help solve equations. When a scale is balanced, both sides are equal.

Try This
Use each balance scale to find two equal weights.

The weight of 1 orange is the same as the weight of _____.

The weight of 1 toy puppy is the same as the weight of

_____.

The weight of 1 box of crayons is the same as the weight of _____ _____.

The weight of 4 baseballs is the same as the weight of _____

_____.

Practice
Find each unknown weight. Draw your answer on the scale with the question mark.

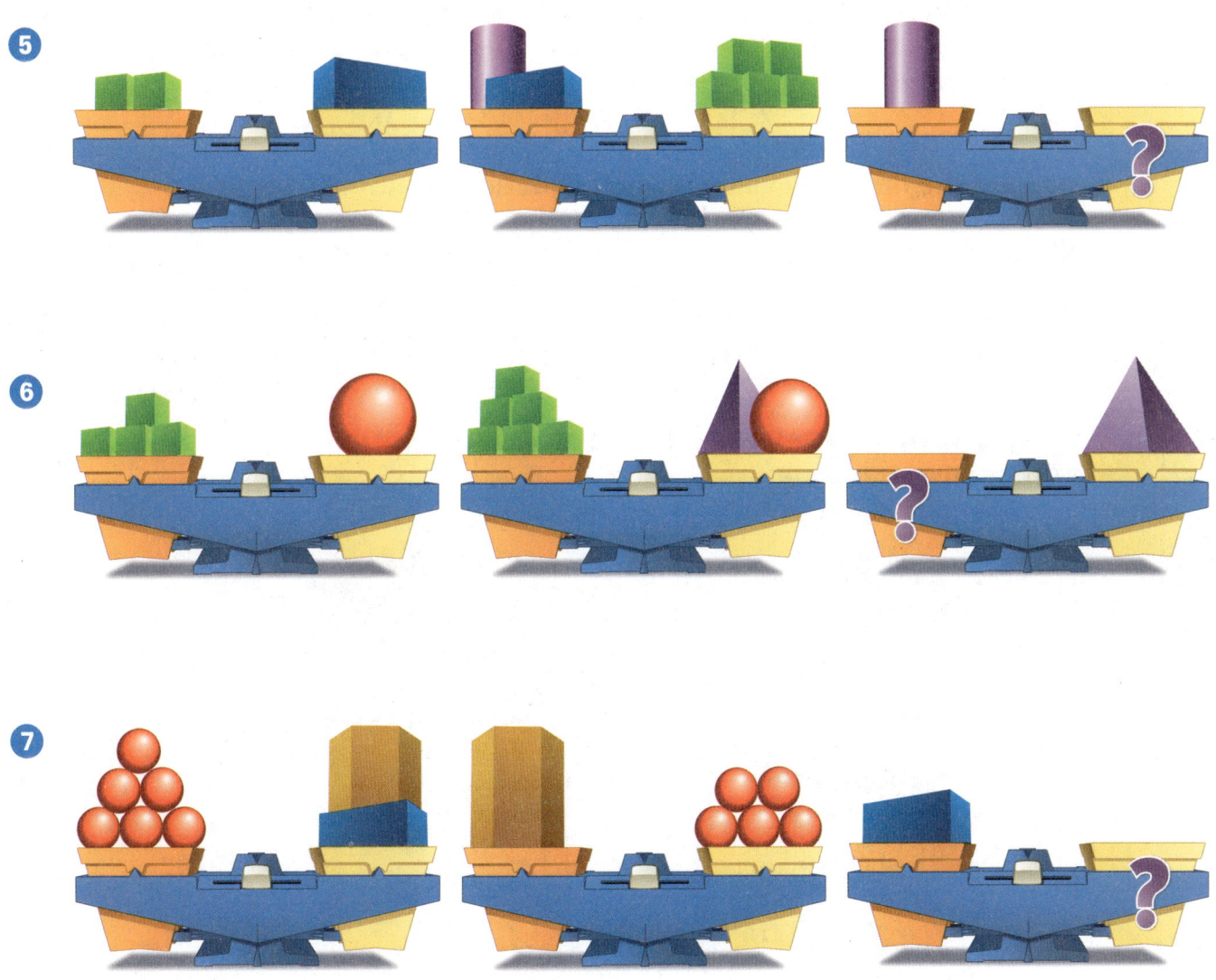

Reflect
What part of a number sentence is represented by the balance scale? Explain.

_____
_____
_____

Variables and Equality • Lesson 4

Week 4
Variables and Equality

Lesson 5 Review

This week you explored equality and unknown values in number sentences. You used shapes to represent missing numbers in an equation. You also related number sentences to weights and balance scales.

Lesson 1 Find the unknown value in each equation.

① $c + 2 = 8$

$c =$ _____

② ☐ + ☐ + ☐ = 9

What is ☐ ?

Lesson 2 Find each unknown weight. Write a number sentence to show your work.

③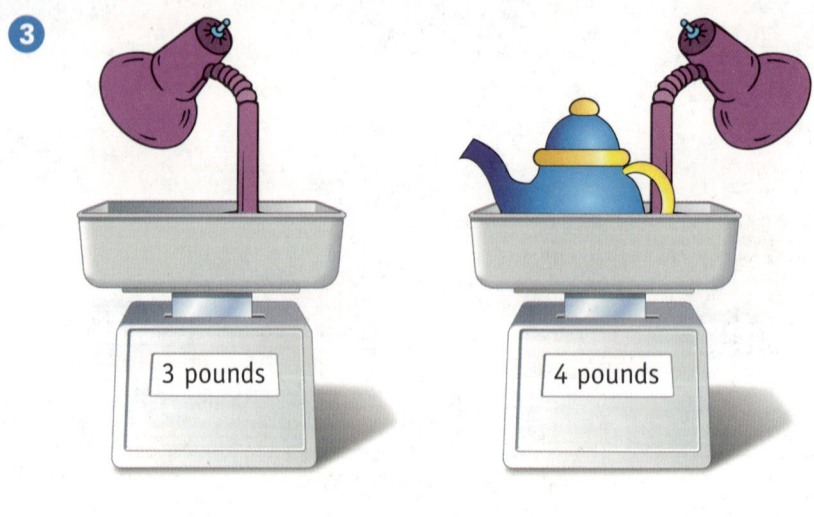

3 pounds   4 pounds

_____

The teapot weighs _____.

Reflect

Explain how to find the values of the unknowns in the number sentence.

☐ + ☐ + ☐ = 21

_____

_____

40 Number Patterns and Relationships • Week 4

Lesson 3

The knife weighs _____.

Lesson 4 Find the unknown weight. Draw your answer on the scale with the question mark.

Reflect
Use shapes from Problem 5 to balance each scale. Draw your answer on the scale with the question mark.

Week 1 Exploring Patterns

Practice

1 Draw the next set in the pattern.

Set 1   Set 2   Set 3   _____

2 Draw the missing set in the pattern.

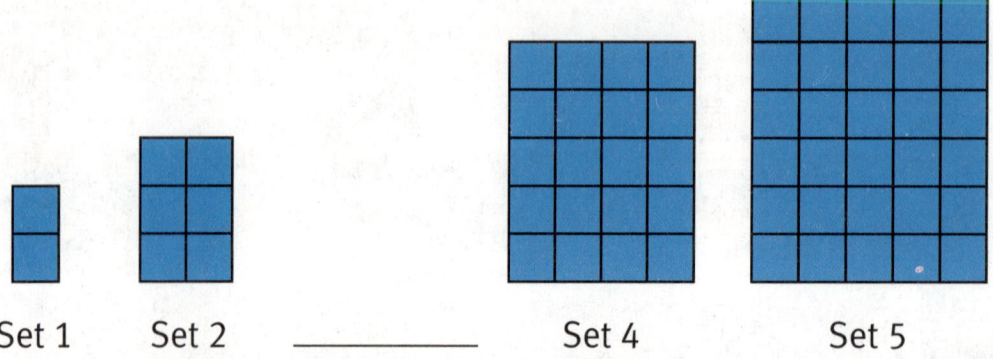

Set 1   Set 2   _____   Set 4   Set 5

3 Tell whether each pattern above is a same-step or a changing-step growing pattern.

_____

_____

4 Design a same-step or changing-step pattern. Show sets 1, 2, and 3, and describe the pattern.

42 Number Patterns and Relationships • Week 1 Practice

Week 2 — Patterns and Relationships

Practice

Complete the table for the pattern shown below.

1

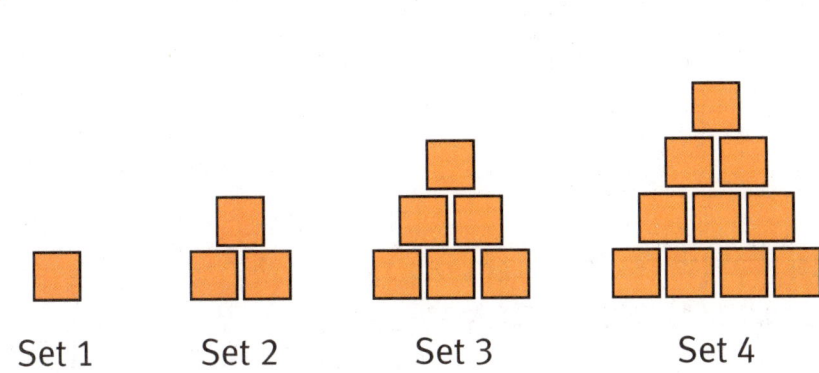

Set	Number of Squares

Complete each input/output table.

2 Candy bars cost 50¢ each.

Input (number of candy bars)	Output (total cost)

3 Jean can ride her bike 20 miles per hour.

Input (number of hours)	Output (number of miles)

Number Patterns and Relationships • Week 2 Practice

Week 3 Patterns and Graphs

Practice

1 The Booster Club sells gourmet cookies for $1.50 each. Complete the input/output table, and graph the pattern.

Input (number of cookies)	Output (total cost)

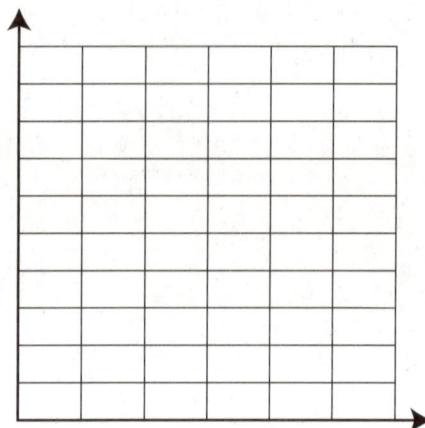

2 Describe the pattern that is shown in the graph.
_____
_____
_____

3 How much would it cost to purchase 10 cookies?
_____

4 What remains the same in the graph?
_____

Week 4

Variables and Equality

Practice

1. $x - 5 = 2$

$x =$ _____

2. $10 - y = 6$

$y =$ _____

3. $z + 5 = 15$

$z =$ _____

4. ☐ + ☐ + ☐ = 9

What is ☐ ?

5. $22 - \triangle = \triangle$

What is △ ?

6. ◯ + ◯ = 12

What is ◯ ?

7. △ + △ = 18

What is △ ?

8. $m - 6 = 2$

$m =$ _____

9. Explain how to find the values of the unknowns in the number sentence

☐ + ☐ + ☐ = 27

_____

_____

_____

_____

Number Patterns and Relationships • Week 4 Practice

Unit 2 Workbook
Level F

SRA
NUMBER WORLDS

Number Patterns and Relationships

featuring Building Blocks Software

Author
Sharon Griffin
*Associate Professor of Education and
Adjunct Associate Professor of Psychology*
Clark University
Worcester, Massachusetts

Building Blocks Authors

Douglas H. Clements
*Professor of Early Childhood
and Mathematics Education*
University at Buffalo
State University of New York, New York

Julie Sarama
Associate Professor of Mathematics Education
University at Buffalo
State University of New York, New York

Contributing Writers
Sherry Booth, *Math Curriculum Developer,* Raleigh, North Carolina
Elizabeth Jimenez, *English Language Learner Consultant,* Pomona, California

Program Reviewers

Jean Delwiche
Almaden Country School
San Jose, California

Cheryl Glorioso
Santa Ana Unified School District
Santa Ana, California

Sharon LaPoint
School District of Indian River County
Vero Beach, Florida

Leigh Lidrbauch
Pasadena Independent School District
Pasadena, Texas

Dave Maresh
Morongo Unified School District
Yucca Valley, California

Mary Mayberry
Mon Valley Education Consortium, AIU 3
Clairton, Pennsylvania

Lauren Parente
Mountain Lakes School District
Mountain Lakes, New Jersey

Juan Regalado
Houston Independent School District
Houston, Texas

M. Kate Thiry
Dublin City School District
Dublin, Ohio

Susan C. Vohrer
Baltimore County Public Schools
Baltimore, Maryland

SRAonline.com

Copyright © 2007 SRA/McGraw-Hill.

All rights reserved. Except as permitted under the United States Copyright Act, no part of this publication may be reproduced or distributed in any form or by any means, or stored in a database or retrieval system, without the prior written permission of the publisher, unless otherwise indicated.

Printed in the United States of America.

Send all inquiries to:
SRA/McGraw-Hill
4400 Easton Commons
Columbus, OH 43219

R53180.01

7 8 9 QPE 12 11 10 09

Photo Credits
2–14 ©PhotoDisc/Getty Images, Inc.; **15** ©Stockbyte/Stockbyte; **16–23** ©PhotoDisc/Getty Images, Inc.; **24** ©Eyewire/Getty Images, Inc.; **25** ©Matt Meadows; **33** ©Stockbyte/Stockbyte

Contents

Number Patterns and Relationships

Week 1 Exploring Patterns 2

Week 2 Patterns and Relationships 12

Week 3 Patterns and Graphs 22

Week 4 Variables and Equality 32

Week 1 Practice 42

Week 2 Practice 43

Week 3 Practice 44

Week 4 Practice 45

Week 1 — Exploring Patterns

Lesson 1

Key Idea

Same-step patterns are patterns in which the amount of change or growth is the same from one set to the next.

Ask the following questions to help you look for a pattern.

- What is changing from one set to the next?
- What stays the same from one set to the next?
- How would you create the next set?

Try This

Describe what changes in each pattern. Sketch and label the next figure in the pattern.

 1

Set 1   Set 2   Set 3   Set 4   _____

_____

_____

2

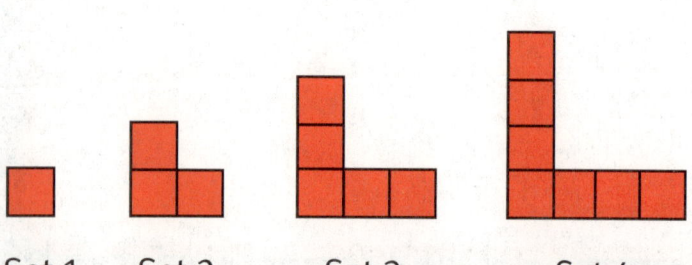

Set 1   Set 2   Set 3   Set 4   _____

_____

_____

2 Number Patterns and Relationships • Week 1

Practice
Sketch and label the next two terms in each pattern.

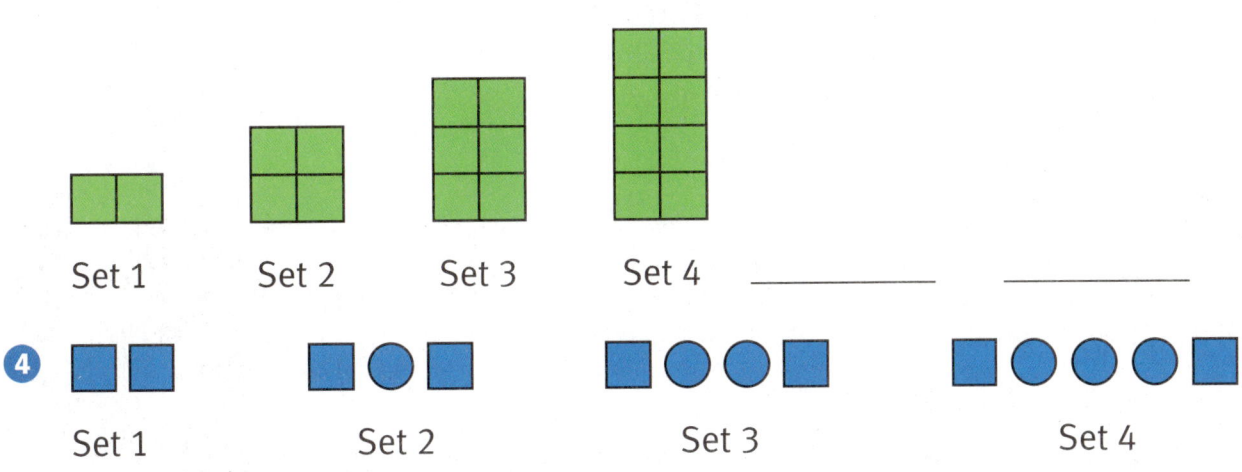

Reflect
Mandy had 5 pennies in her piggy bank on Monday. Each day she put 10 more pennies into the bank. How much money will she have in the piggy bank on Friday? Explain.

Monday	Tuesday	Wednesday	Thursday	Friday
5¢	15¢	25¢	35¢	_____

_____
_____
_____
_____

Exploring Patterns • Lesson 1

Week 1 **Exploring Patterns**

Lesson 2

Key Idea

Changing-step patterns are patterns in which the amount of change or growth changes in a regular and predictable way from one set to the next.

Ask the following questions to help you identify a changing-step pattern:

- What changes from one set to the next?
- What stays the same from one set to the next?
- How would you create the next set?

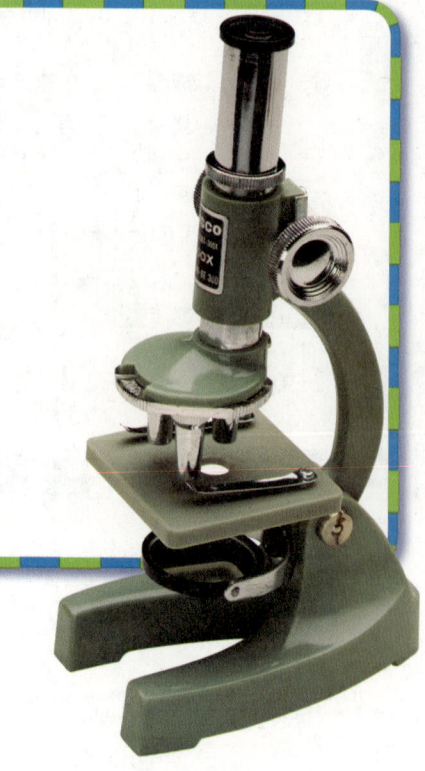

Try This

Describe the changes in each pattern. Then sketch and label the next set in the pattern.

1

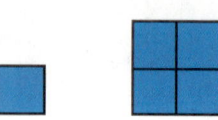

Set 1   Set 2   Set 3   Set 4   _____

_____
_____

2

Set 1   Set 2   Set 3   _____

_____
_____

Practice
Sketch and label the next term in the pattern.

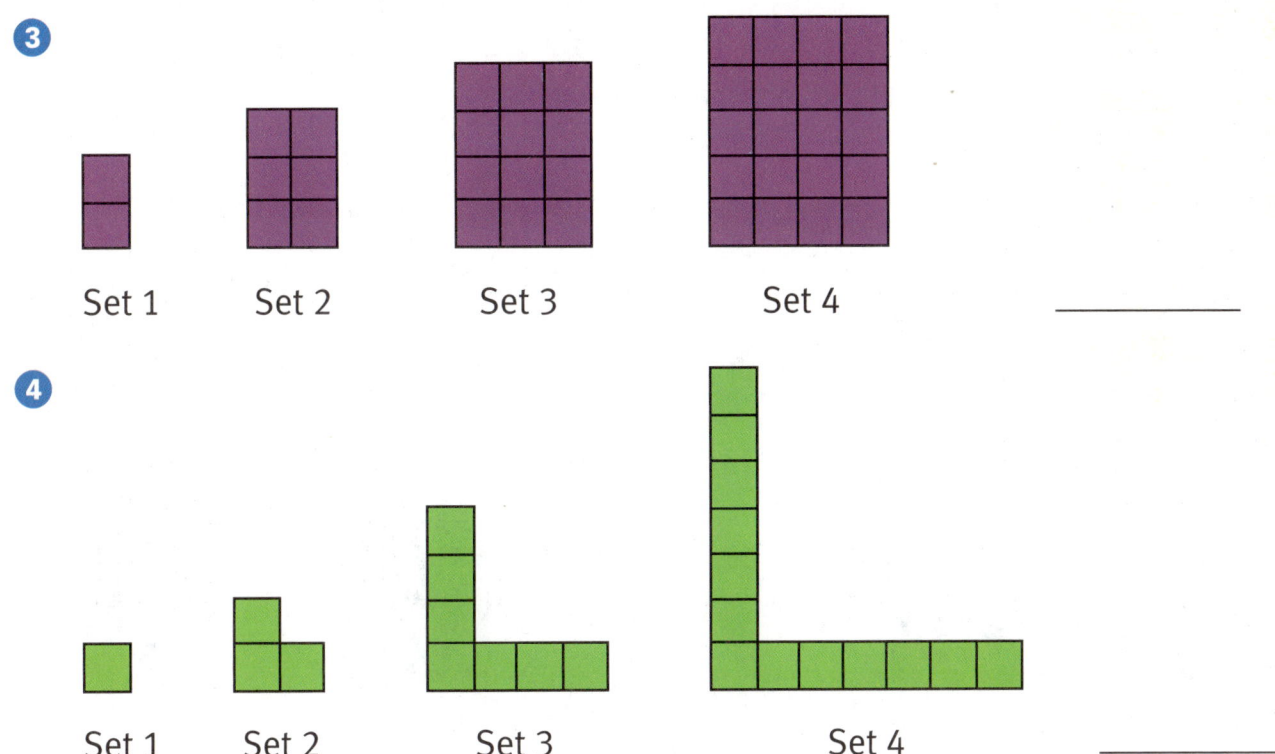

❸ Set 1, Set 2, Set 3, Set 4, _____

❹ Set 1, Set 2, Set 3, Set 4, _____

Reflect
Look for a pattern in the table. If the pattern continues, how many chin-ups will Jarrod do on Friday? Explain your answer.

Jarrod's Chin-up Chart				
Monday	Tuesday	Wednesday	Thursday	Friday
8 chin-ups	10 chin-ups	13 chin-ups	17 chin-ups	

_____
_____
_____

Exploring Patterns • Lesson 2

Week 1 Exploring Patterns

Lesson 3

Key Idea
You can use clues from the sets given in a pattern to identify missing sets.

Try This
Look for the changes in each pattern. Then draw and label the missing set.

1 Set 1   Set 2   _____   Set 4   Set 5

2 Set 1   Set 2   _____   Set 4

3 Set 1   _____   Set 3   Set 4   Set 5

6 Number Patterns and Relationships • Week 1

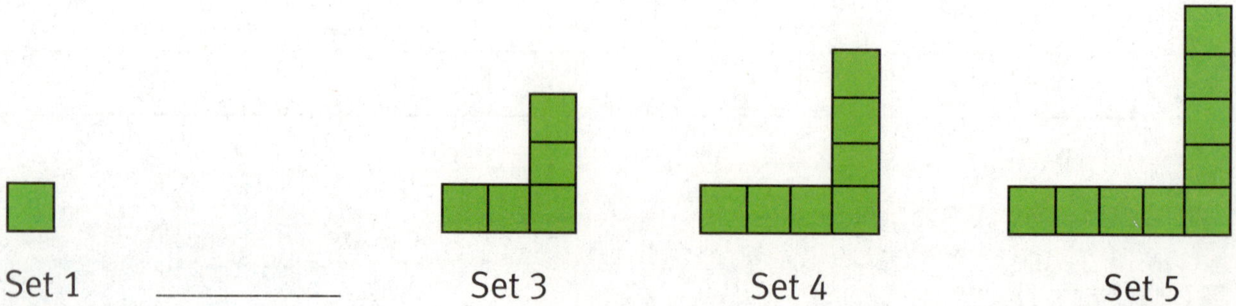

Practice
Look for the changes in each pattern. Tell whether it is a same-step growing pattern or a changing-step growing pattern. Then draw and label the missing set.

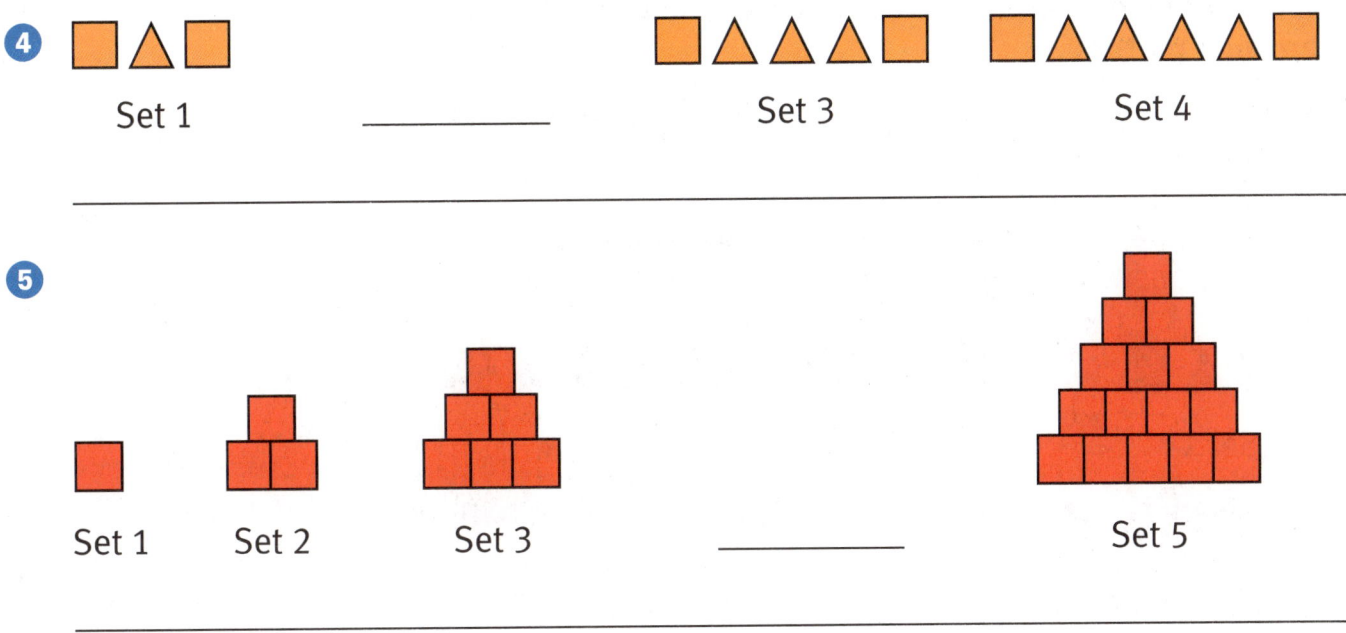

Reflect
Do you think it is easier to find a missing set in a same-step or a changing-step growing pattern? Explain.

_____
_____
_____
_____

Week 1 — Exploring Patterns

Lesson 4

> **Key Idea**
> You can use pattern blocks to create your own growing patterns.

Try This

Draw and label the next set for each pattern. Use the pattern to answer the questions.

1

Set 1 Set 2 Set 3 _____

a. Is this a same-step or a changing-step growing pattern?

_____

b. How is the pattern changing from set to set?

_____

2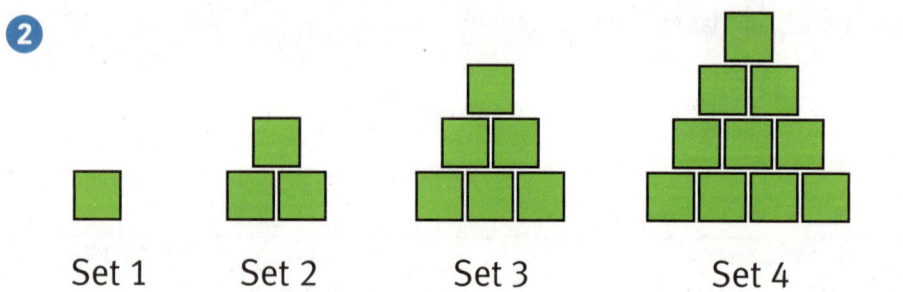

Set 1   Set 2   Set 3   Set 4   _____

a. Is this a same-step or a changing-step growing pattern?

_____

b. How is the pattern changing from set to set?

_____

_____

Practice

Draw your own pattern in the space below. Exchange your pattern with a partner, and have your partner answer each question.

③ Is this a same-step or a changing-step growing pattern?

_____

④ How is the pattern changing from set to set?

_____

⑤ What would the next set of the pattern look like? Describe it.

_____

Reflect

Use different shapes to create the first few sets of a pattern. Explain how the pattern changes from set to set.

Exploring Patterns • Lesson 4

Week 1 — Exploring Patterns

Lesson 5 Review

This week you explored patterns. You examined same-step patterns and changing-step patterns.

Lesson 1 Sketch and label the next set in the pattern.

1

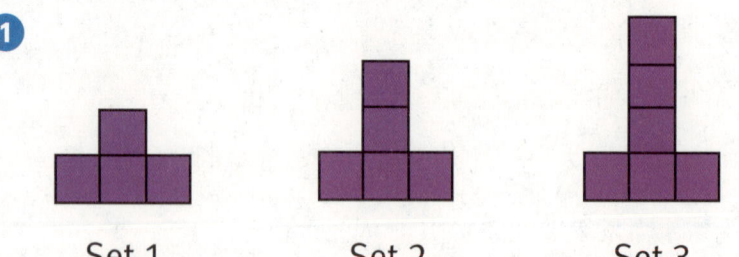

Set 1   Set 2   Set 3   _____

2

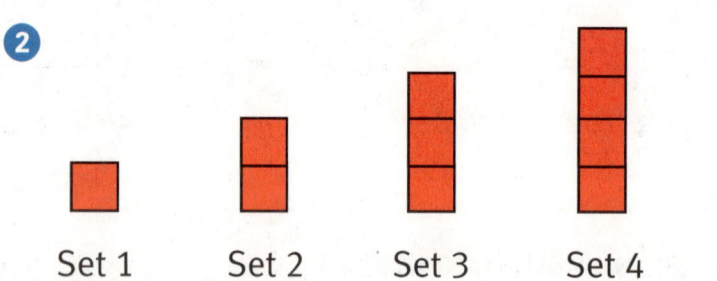

Set 1   Set 2   Set 3   Set 4   _____

Lesson 2 Sketch and label the next set in the pattern.

3

Set 1   Set 2   Set 3   _____

Reflect

Tell whether each pattern above is a same-step or a changing-step growing pattern.

_____

_____

_____

10   Number Patterns and Relationships • Week 1

Lesson 3 **Sketch and label the missing set in the pattern.**

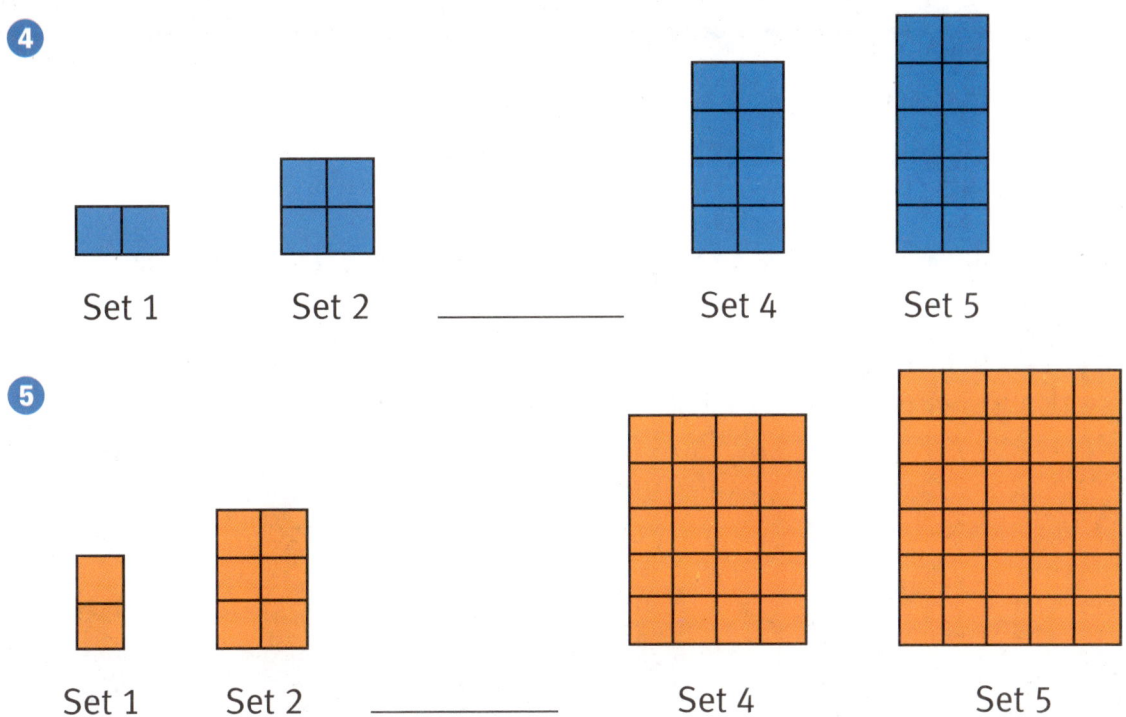

④ Set 1   Set 2   _____   Set 4   Set 5

⑤ Set 1   Set 2   _____   Set 4   Set 5

Lesson 4 ⑥ Use pattern blocks to design Set 1 of a pattern. Then show Sets 2 and 3.

Reflect
Describe how to find the missing set. Then draw and label the missing set.

Set 1   Set 2   _____   Set 4

_____

_____

_____

_____

Week 2 — **Patterns and Relationships**

Lesson 1

> **Key Idea**
> You can use numbers to represent the growth in patterns.

Try This
Use the growing pattern below to answer each question.

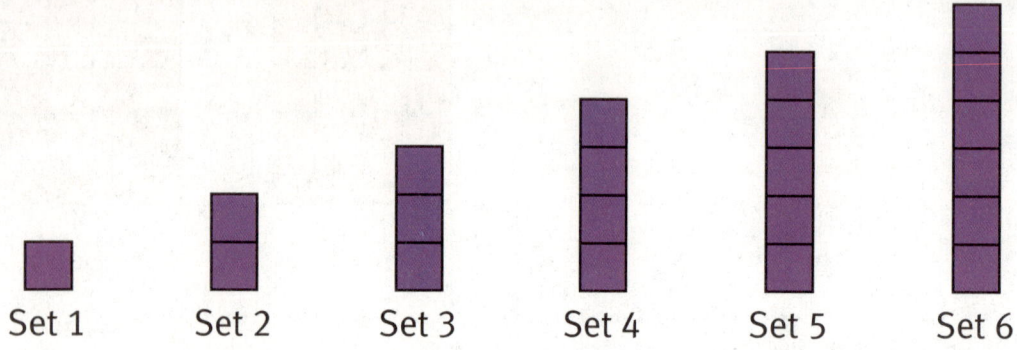

1. Use the pattern to complete the table below.

Set	Number of Squares
1	
2	
3	
4	
5	
6	

2. What patterns do you notice in the table between the set number and the number of squares?

_____

_____

3. Use the pattern from Problem 2 to predict the number of squares in the tenth set without building or drawing all the sets up to the tenth.

_____

Practice
Use the pattern to complete the table.

④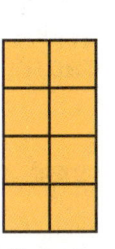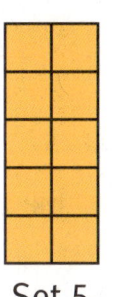

Set 1   Set 2   Set 3   Set 4   Set 5

Set	Number of Squares
1	
2	
3	
4	
5	

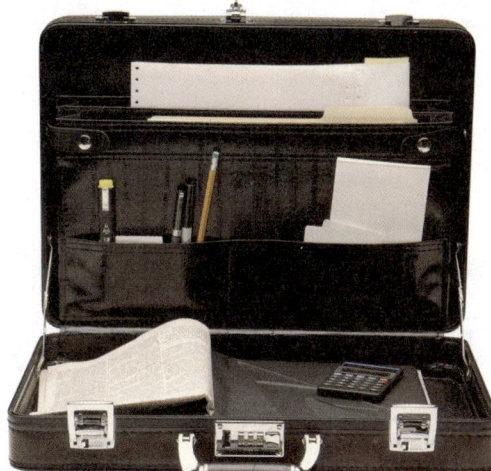

⑤ What patterns do you notice in the table above between the set number and the number of squares?

_____

⑥ Use the pattern from Problem 5 to predict the number of squares in the tenth set without building or drawing all of the sets up to the tenth.

_____

Reflect
When you use a pattern table to help you create a pattern with shapes, what is the smallest number of sets you need to figure out the pattern? Explain.

_____

_____

_____

Patterns and Relationships • Lesson 1

Week 2 — Patterns and Relationships

Lesson 2

Key Idea
You can use numbers to represent the growth in patterns.

Try This
Use the growing pattern below to answer each question.

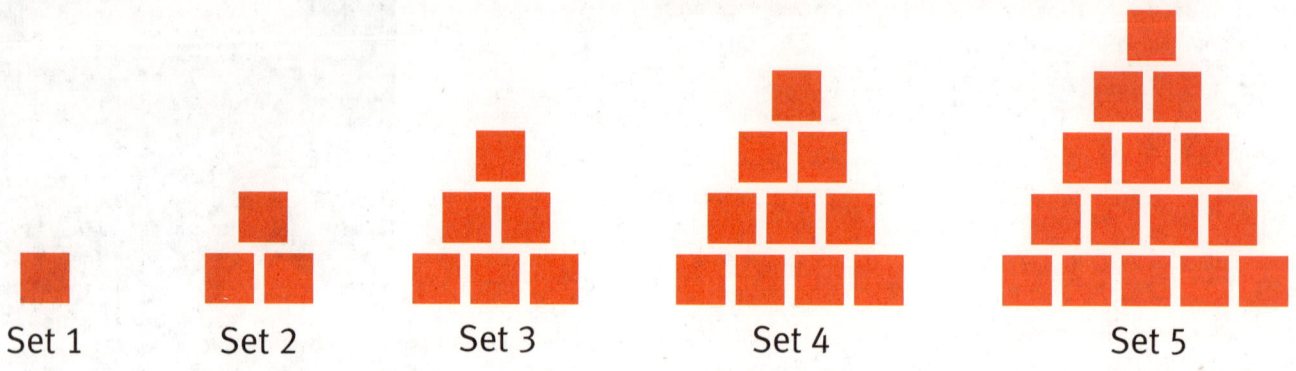

Set 1   Set 2   Set 3   Set 4   Set 5

1. Is this a same-step or a changing-step growth pattern?
   _____

2. Describe the growth shown in the pattern.
   _____

3. How many squares were used to create each set?
   _____

4. Use your answers to complete the table below.

Set	Number of Squares
1	
2	
3	
4	
5	

14 Number Patterns and Relationships • Week 2

Practice
Use the pattern to complete the table.

5

  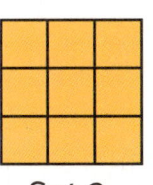  

Set 1   Set 2   Set 3   Set 4   Set 5

Set	Number of Squares
1	
2	
3	
4	
5	

6 What patterns do you notice in the table above between the set number and the number of squares?

_____

7 Use the pattern to predict the number of squares in the tenth set without building or drawing all of the sets up to the tenth.

_____

Reflect
What do you notice about the table for a changing-step pattern? How does it compare with the table for a same-step pattern?

_____
_____
_____

Week 2 — Lesson 3
Patterns and Relationships

Key Idea
The pattern tables are examples of input/output tables.

For each input value (the set number), there is a certain output value (number of squares).

Input/output tables can also be used to model real-world situations.

Try This
Thomas mows lawns in his neighborhood to earn money. He earns $8 for each lawn he mows. Use the input/output table to answer each question.

Input (lawns mowed)	Output (money earned)
1	$8
2	$16
3	$24
4	$32
5	$40

1 How much money would Thomas earn if he mowed 6 lawns? Explain how you found your answer.

_____

_____

2 Write a mathematical rule for determining the amount of money Thomas will earn for any given number of lawns mowed.

_____

Practice
Complete the table and answer each question.

3 Movie tickets cost $6 each. Complete the input/output table.

Input (number of tickets)	Output (total cost)
1	$6
2	
3	
4	
5	

4 How much would it cost to buy 4 movie tickets?

_____

5 Write a mathematical rule for determining the total cost of tickets for any given number of tickets.

_____

_____

6 For every hour Lisa drives, she uses 2 gallons of gasoline. Her gas tank holds 18 gallons when it is full. Complete the input/output table.

Input (hours of driving)	Output (gas remaining in her tank)
1	16 gallons
2	14 gallons
3	
4	
5	

7 How much gasoline is in Lisa's tank after 5 hours of driving?

_____

8 Write a mathematical rule for determining the amount of gas remaining for any given number of hours driven.

_____

_____

Reflect
What is different about the lawn mowing input/output table and the gasoline input/output table?

_____

_____

_____

Patterns and Relationships • Lesson 3 **17**

Week 2 — Patterns and Relationships

Lesson 4

Key Idea
You can use input/output tables to help you make choices.

Try This
Anna's neighbors have hired her to pet sit their dog for seven days. They have offered two different options for being paid.

- **Option 1:** Anna receives $10 for the first day and an additional $2 per day after the first day.

- **Option 2:** Anna receives $1 for the first day. Every day after the first day she receives an additional amount that is $1 more than the previous day.

❶ Complete the input/output tables for each option.

Option 1	
Day	Total Amount Earned
1	$10
2	$12
3	
4	
5	
6	
7	

Option 2	
Day	Total Amount Earned
1	$1
2	$3
3	$6
4	
5	
6	
7	

❷ If Anna chooses Option 1, how much money will she be paid? If Anna chooses Option 2, how much money will she be paid?

Practice

Jim was hired to do yard work for his neighbor. The neighbor expects the work to last 5 days, but it could last 7 days. Payment options are as follows:

- **Option 1:** Jim receives $12 for the first day and $2 per day after the first day.

- **Option 2:** Jim receives $2 for the first two days. Every day thereafter he receives an amount that is $1 more than the previous day.

Complete the tables to find Jim's total earnings for the week for each option.

❸ Complete the table for each option.

Option 1		
Day	Amount Earned for the Day	Total Earnings
1	$12	
2		
3		
4		
5		
6		
7		

Option 2		
Day	Amount Earned for the Day	Total Earnings
1	$2	
2		
3		
4		
5		
6		
7		

❹ How much will Jim earn for 7 days if he chooses Option 1? _____

❺ How much will Jim earn for 7 days if he chooses Option 2? _____

❻ If the job is for only 5 days, which option will pay better? _____

Reflect

What kind of growth pattern is shown by Option 1? What kind of growth pattern is represented by Option 2?

Week 2: Patterns and Relationships

Lesson 5 Review

This week you explored patterns and relationships. You looked at how visual patterns can be related to number patterns. You also learned about input/output tables and solving problems.

Lessons 1 and 2

Complete the table for the pattern shown below.

1

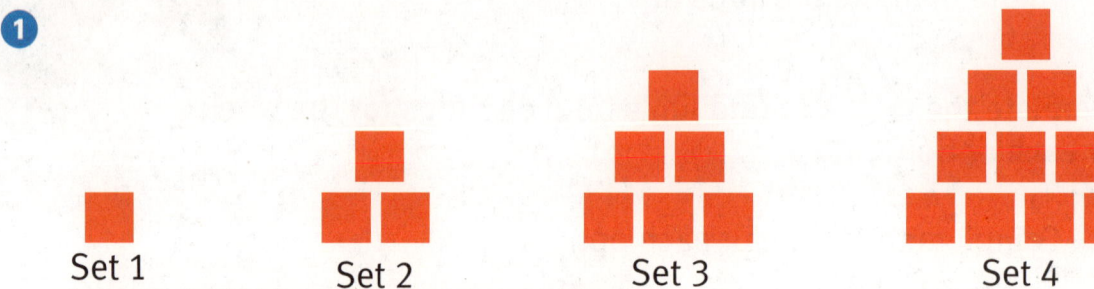

Set 1   Set 2   Set 3   Set 4

Set	Number of Cubes
1	
2	
3	
4	

Reflect

How is the pattern changing?

_____

20 Number Patterns and Relationships • Week 2

Lesson 3 Complete the input/output table.

2 The bookstore sells pencils for 15¢ each.

Input (number of pencils)	Output (total cost)
1	15¢
2	
3	
4	

Lesson 4 A bathtub holds 60 gallons of water. When the drain plug is pulled, 12 gallons drain from the tub each minute.

3 How long does it take for the tub to fully drain?

_____

Input (number of minutes)	Output (water remaining in the tub)
0	60 gallons
1	
2	
3	
4	
5	

Reflect

How many gallons of water are left in the tub 1 minute after the plug is pulled? How many gallons of water are left in the tub 3 minutes after the plug is pulled? Is this an example of same-step pattern or a changing-step pattern?

_____

_____

_____

Week 3

Patterns and Graphs

Lesson 1

Key Idea
Patterns can be represented with pictures, rules, and tables. They can also be represented with graphs.

Try This

Below is a graph that shows how far Lisa can drive, depending on the number of gallons of gasoline in the car's tank.

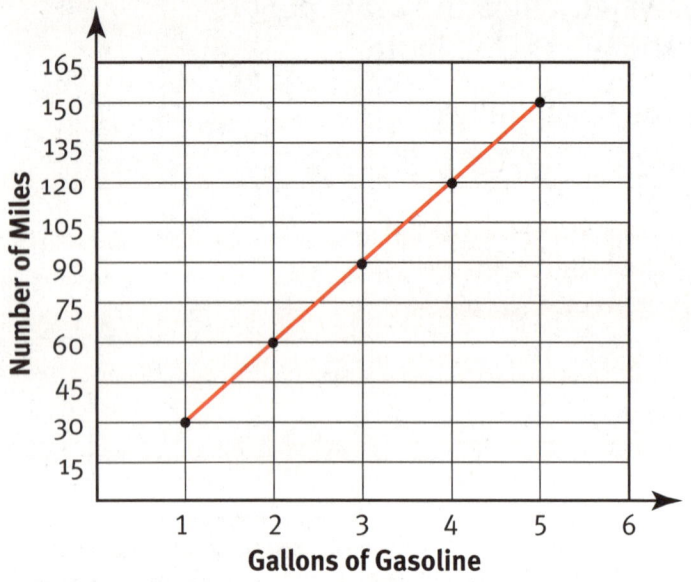

1 Which axis represents the number of miles Lisa can drive?

_____

2 Which axis represents the amount of gasoline in Lisa's car?

_____

3 As the number of gallons of gasoline increases, what happens to the distance that Lisa can drive? Is this increase a same-step increasing pattern or a changing-step pattern?

_____

_____

4 Describe the pattern shown in the graph.

_____

_____

_____

Practice

The graph shows the money Thomas made mowing lawns.

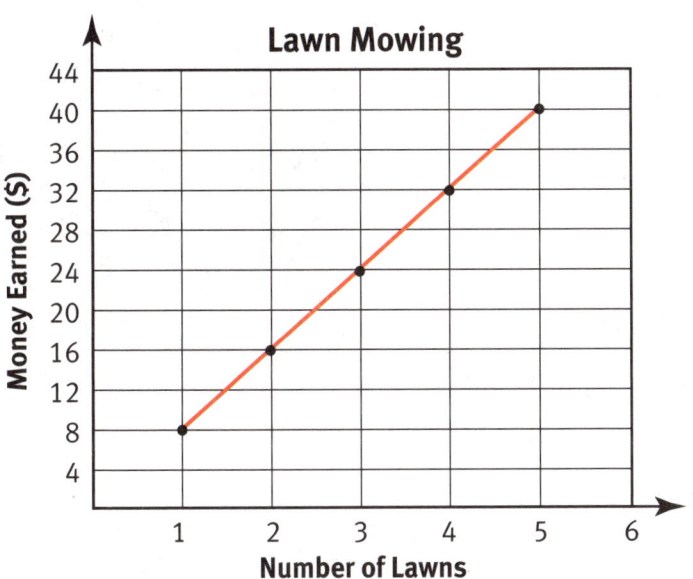

5 The data points are connected to help you see the trend of the data. As the number of lawns mowed increases, what happens to the amount of money earned?

_____

6 By connecting the data points, you are showing that the data is continuous. Should these data points be connected? Explain your answer.

_____
_____
_____
_____

Reflect
Create an input/output table, using the information above.

Patterns and Graphs • Lesson 1   23

Week 3

Patterns and Graphs

Lesson 2

Key Idea
When creating a graph, be sure to label the axes and give it a title.

Try This
Follow the steps to create a graph of the pattern.

Input (movie tickets)	Output (total cost)
1	$6
2	$12
3	$18
4	$24
5	$30

Step 1 Label the horizontal axis and the vertical axis.

Step 2 Plot a point for each pair of numbers in the table.

Step 3 Give your graph a title.

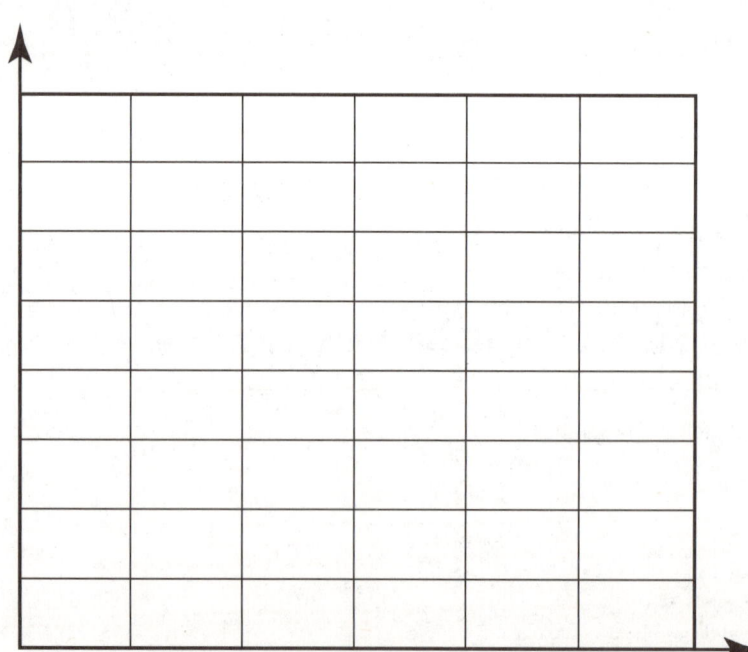

24 Number Patterns and Relationships • Week 3

Practice

Mandy is babysitting for her neighbors. Graph the pattern shown in the input/output table.

Input (number of hours)	Output (money earned)
1	$5
2	$10
3	$15
4	$20
5	$25

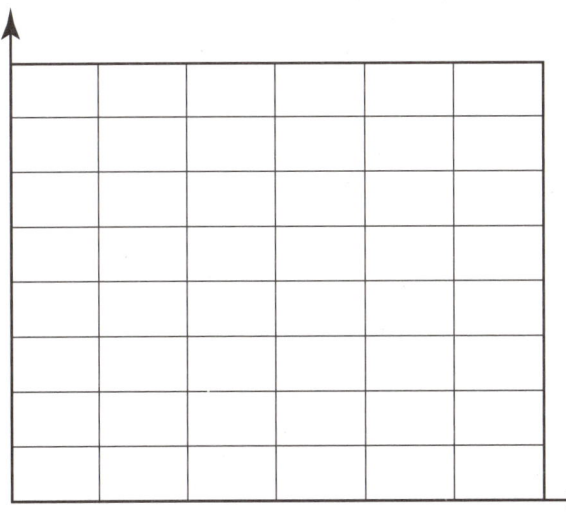

1 Describe the pattern shown in the graph.

_____

2 How much does Mandy earn for babysitting 6 hours?

_____

Reflect

Can you determine the rule for a pattern by just looking at the graph? Explain and give an example.

_____

_____

_____

Patterns and Graphs • Lesson 2

Week 3 — Patterns and Graphs

Lesson 3

> **Key Idea**
> You can use graphs to compare two related patterns.

Try This

Create a graph for each input/output table.
Answer each question.

Milk Cartons Sold	
Input (day)	Output (milk sold for the week)
Monday	25 cartons
Tuesday	50 cartons
Wednesday	75 cartons
Thursday	100 cartons
Friday	125 cartons

Milk Cartons Left	
Input (day)	Output (milk cartons left in the cafeteria)
Monday	125 cartons
Tuesday	100 cartons
Wednesday	75 cartons
Thursday	50 cartons
Friday	25 cartons

① Which arrow line or axis represents the day of the week?
Which represents the number of milk cartons left?

___

26 Number Patterns and Relationships • Week 3

Practice
Use your graphs from Try This to answer each question.

2 Describe the pattern shown in the first graph.
___

3 Describe the pattern shown in the second graph.
___

4 Which of the graphs shows a growing pattern?
___

5 Are these graphs same-step patterns or changing-step patterns?
___

6 What stays the same in the first graph? What changes?
___

7 What stays the same in the second graph? What changes?
___

8 How are the two graphs related?
___
___

Reflect
Can you show both patterns on the same graph? Explain.
___
___
___

Patterns and Graphs • Lesson 3

Week 3 — Patterns and Graphs

Lesson 4

Key Idea
You can use graphs to tell a story or make an informed decision.

Try This
Choose the story that belongs with each graph.

Story A Melissa rode her bike for 40 minutes. The table shows the distance she traveled.

Time	10 minutes	20 minutes	30 minutes	40 minutes
Distance	3 miles	6 miles	9 miles	12 miles

Story B Mrs. Swanson walked her dog for 40 minutes. The table shows the number of blocks she covered.

Time	10 minutes	20 minutes	30 minutes	40 minutes
Distance	6 blocks	12 blocks	18 blocks	24 blocks

1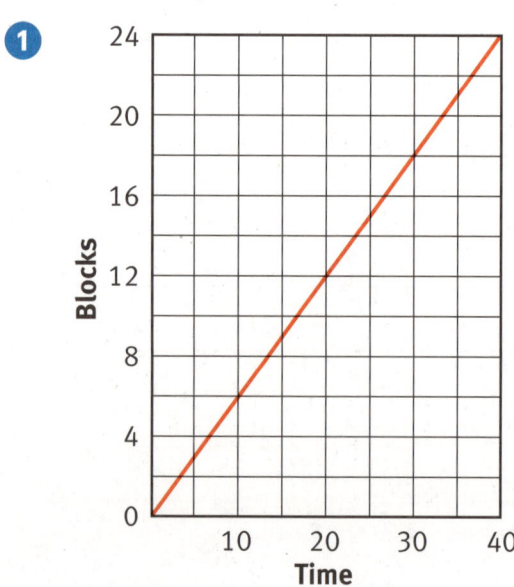

_____

2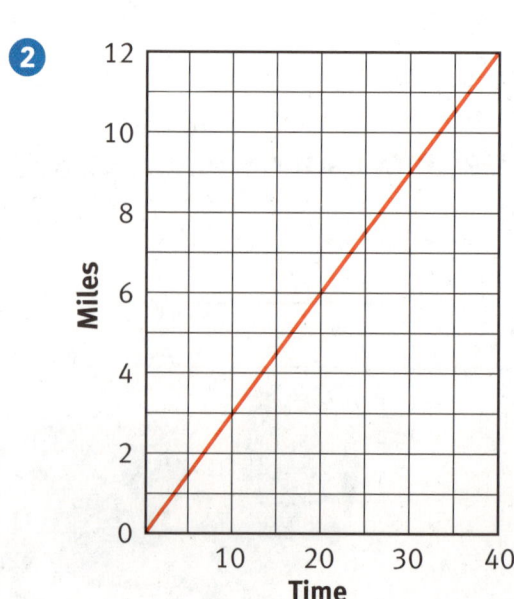

_____

28 Number Patterns and Relationships • Week 3

Practice
Create a graph for the data in the table.

3.

Number of Books	1	2	4	6
New Vocabulary Words	2	4	8	12

Reflect
Create a table that compares the outside temperature throughout the morning and afternoon of a winter day. Graph the data in the table.

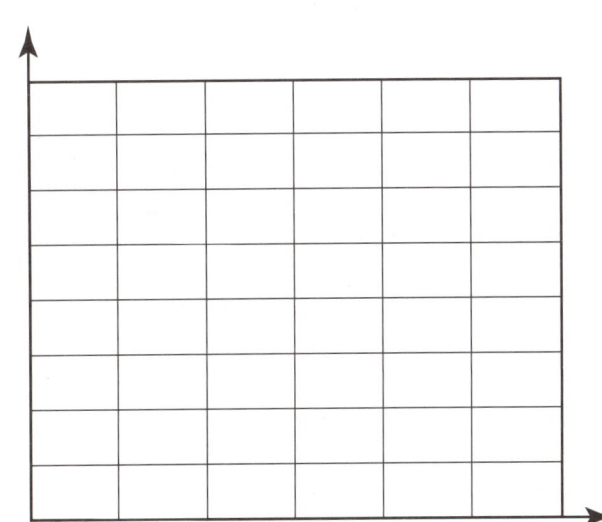

Patterns and Graphs • Lesson 4   29

Week 3 — Patterns and Graphs

Lesson 5 Review

This week you explored how patterns look in graphs. You used input/output tables and stories to create graphs. You also used graphs to answer questions about the pattern and data.

Lessons 1 and 2

The bookstore sells school sweatshirts for $10 each. Graph the pattern shown in the input/output table.

Input (number of sweatshirts)	Output (total cost)
1	$10
2	$20
3	$30
4	$40
5	$50

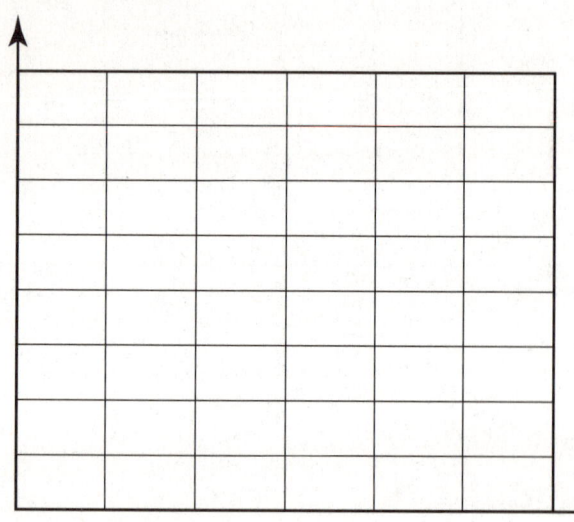

1. Describe the pattern that is shown in the graph.
   _____
   _____

2. How much would it cost to purchase eight sweatshirts?
   _____

Reflect
What is staying the same in the graph?
_____

30 Number Patterns and Relationships • Week 3

Lesson 3 Graph the pattern shown in the input/output table.

Input (number of cars)	Output (total passengers)
1	4
2	8
3	12
4	16
5	20

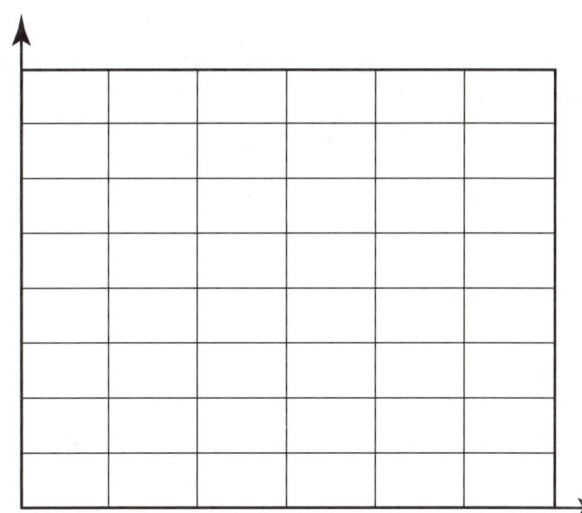

Lesson 4 Use the graph to answer each question.

❸ Which story matches the graph? Circle A or B.

A. A shoe store sold 20 pairs of shoes on Monday. The store sold no shoes on Tuesday because it was closed. On Wednesday and Thursday, 30 pairs of shoes were sold.

B. A hot air balloon rose to 250 feet. It stayed there for a while and then rose to 500 feet. After a little while longer, the balloon began its descent.

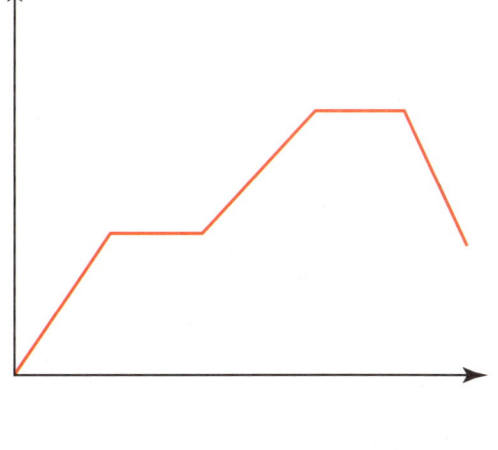

Reflect
What label would you put on the horizontal axis? What label would you put on the vertical axis?

_____

Week 4: Variables and Equality

Lesson 1

Key Idea

An **equation** is a number sentence which states that two mathematical expressions are equal.

$2 + 3 = 5$ $11 - 4 = 7$ $6 - 2 = 3 + 1$

Sometimes equations have unknown values. You can show unknown values with pictures, boxes, or letters.

$4 + \square = 12$ $b - 9 = 5$

Try This

Find the unknown value in each equation. Substitute values into the equation until you have a true number sentence.

1. $\square + 6 = 8$
   What is $\square$?

2. $\triangle + 1 = 7$
   What is $\triangle$?

3. $4 + \triangle = 8$
   What is $\triangle$?

4. $5 - \bigcirc = 2$
   What is $\bigcirc$?

5. $a + 8 = 10$
   $a = \underline{}$

6. $12 - 7 = z$
   $z = \underline{}$

Practice

Find the unknown value in each equation. The same shapes represent the same value.

7. $\square + \square = 10$
   What is $\square$?

8. $\bigcirc + \bigcirc = 2$
   What is $\bigcirc$?

9. △ + △ = 6

What is △ ?

10. ◯ + ◯ + ◯ = 15

What is ◯ ?

Find the unknown value in each equation.

11. 9 − ☐ = 3

What is ☐ ?

12. 8 − n = 6

n = _____

13. t − 7 = 2

t = _____

14. 14 + ◯ = 19

What is ◯ ?

15. x + 4 = 12

x = _____

16. 17 − y = 12

y = _____

Reflect

What values of ☐ and △ make a true number sentence? Is there more than one correct answer? Explain.

☐ − △ = 3

_____

_____

_____

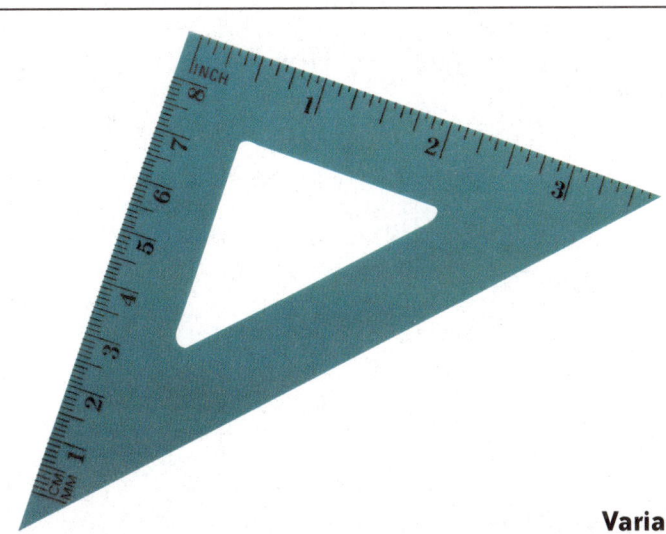

Variables and Equality • Lesson 1

Week 4: Variables and Equality

Lesson 2

> **Key Idea**
> You can use the idea of weights to help solve equations.

Try This
Answer each question to find the weight of the toy car.

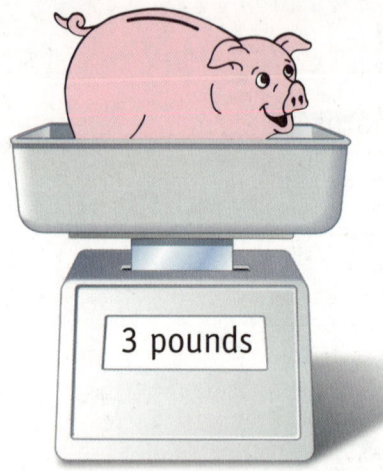
3 pounds

8 pounds

① How much does the piggy bank weigh?

_____

② How much do the piggy bank and toy car weigh altogether?

_____

③ Fill in the blanks below to help you find the weight of the toy car.

Piggy bank = 3

Piggy bank + toy car = 8

_____ + toy car = 8

_____ + _____ = 8

Toy car = _____

The toy car weighs _____.

Practice

Find each unknown weight. Write a number sentence to show your work.

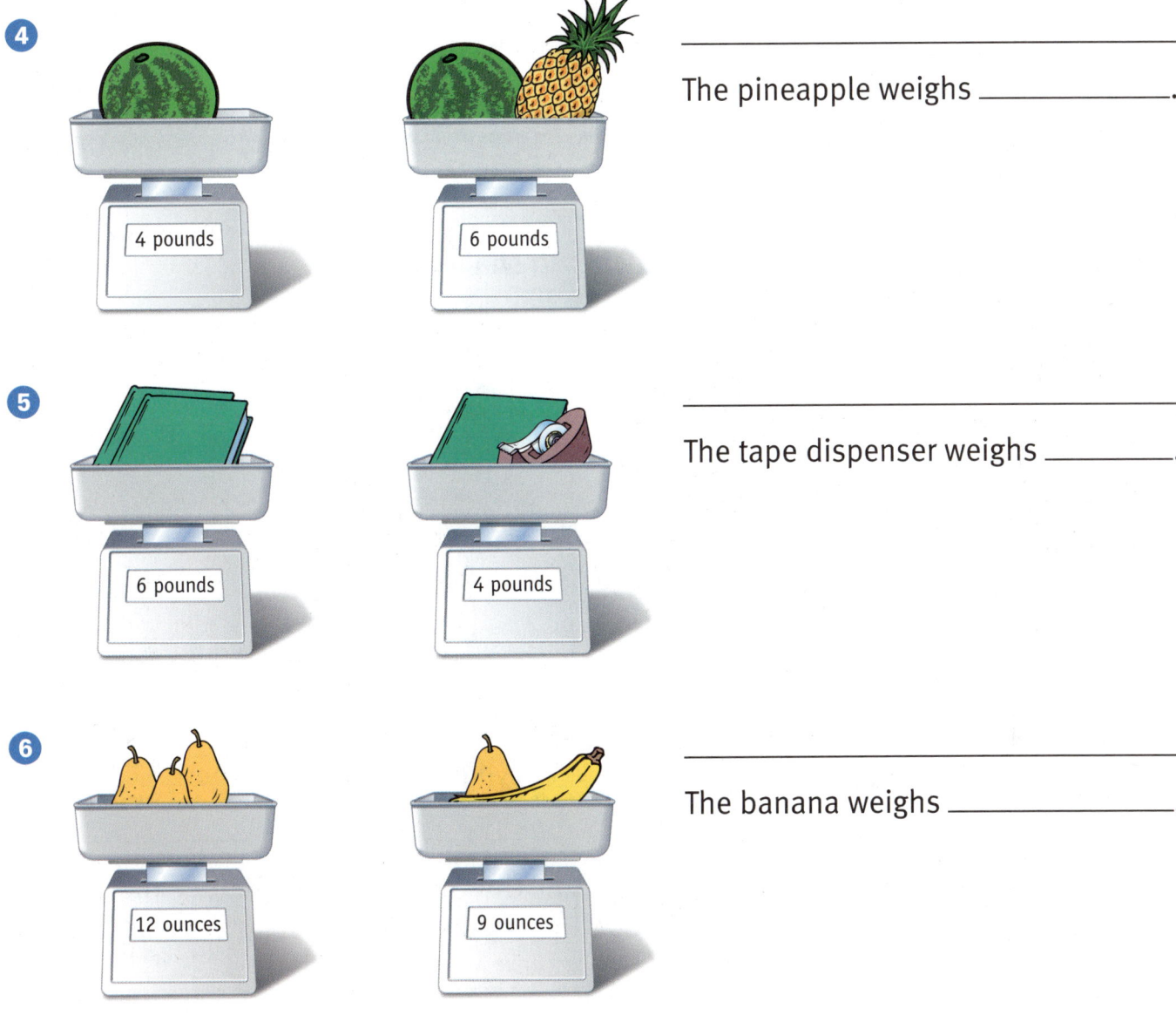

④ The pineapple weighs _____.

⑤ The tape dispenser weighs _____.

⑥ The banana weighs _____.

Reflect

How did you decide how much one pear weighs in Problem 6? Explain.

Variables and Equality • Lesson 2

Week 4

Variables and Equality

Lesson 3

Key Idea
Use reasoning to solve more challenging problems involving weights.

Try This
Answer each question to find the weight of each shape.

1. How much does the pyramid weigh?
_____

2. How much do the pyramid and cylinder weigh altogether?
_____

3. How much does the cylinder weigh?
_____

4. How much do the cylinder and cube weigh altogether?
_____

5. How much does the cube weigh?
_____

Practice
Find each unknown weight.

6 The shoe weighs _____.

7 The basket weighs _____.

8 The plant weighs _____.

9 Each tennis ball weighs _____.

10 Each baseball weighs _____.

11 The volleyball weighs _____.

Reflect
Would you be able to find the weight of the volleyball in Problem 7 if you had only the first and third scales? Explain your answer.

Variables and Equality • Lesson 3   37

Week 4 — Variables and Equality

Lesson 4

> **Key Idea**
> Balance scales can be used to help solve equations. When a scale is balanced, both sides are equal.

Try This

Use each balance scale to find two equal weights.

1

The weight of 1 orange is the same as the weight of _____.

2

The weight of 1 toy puppy is the same as the weight of _____.

3

The weight of 1 box of crayons is the same as the weight of _____ _____.

4

The weight of 4 baseballs is the same as the weight of _____ _____.

Practice

Find each unknown weight. Draw your answer on the scale with the question mark.

5.

6.

7.

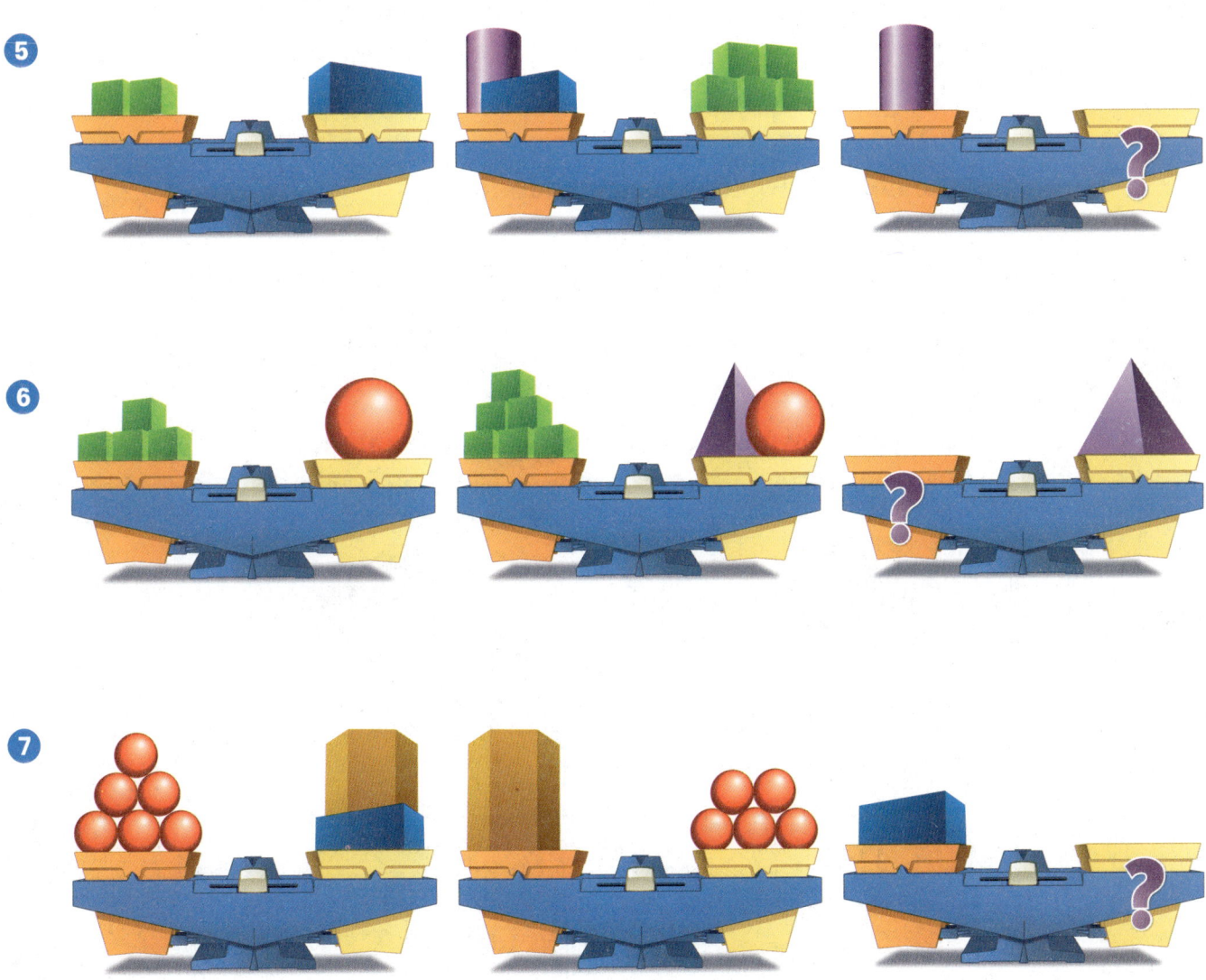

Reflect

What part of a number sentence is represented by the balance scale? Explain.

_____
_____
_____

Variables and Equality • Lesson 4

Week 4 — Variables and Equality

Lesson 5 Review

This week you explored equality and unknown values in number sentences. You used shapes to represent missing numbers in an equation. You also related number sentences to weights and balance scales.

Lesson 1 Find the unknown value in each equation.

① $c + 2 = 8$

$c = $ _____

② $\square + \square + \square = 9$

What is $\square$?

Lesson 2 Find each unknown weight. Write a number sentence to show your work.

③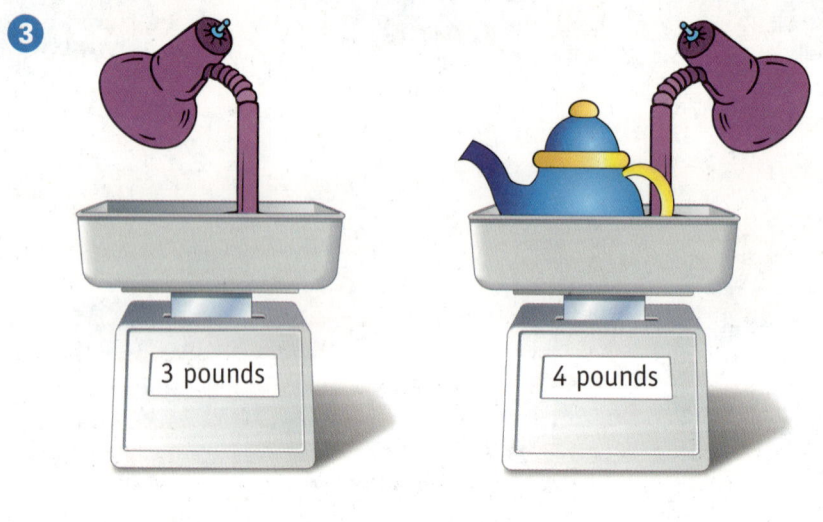

3 pounds   4 pounds

_____

The teapot weighs _____.

Reflect
Explain how to find the values of the unknowns in the number sentence.

$\square + \square + \square = 21$

_____

_____

40 Number Patterns and Relationships • Week 4

Lesson 3 ④

The knife weighs _____.

Lesson 4 Find the unknown weight. Draw your answer on the scale with the question mark.

⑤

Reflect

Use shapes from Problem 5 to balance each scale. Draw your answer on the scale with the question mark.

Week 1 Exploring Patterns

Practice

1 Draw the next set in the pattern.

Set 1   Set 2   Set 3   _____

2 Draw the missing set in the pattern.

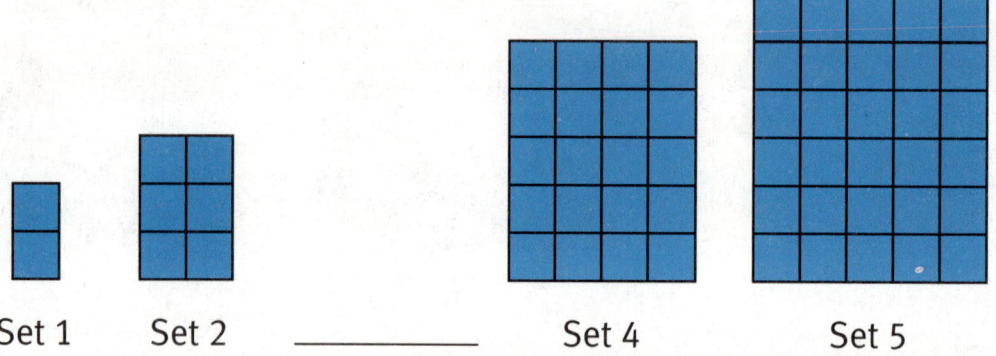

Set 1   Set 2   _____   Set 4   Set 5

3 Tell whether each pattern above is a same-step or a changing-step growing pattern.

_____

_____

4 Design a same-step or changing-step pattern. Show sets 1, 2, and 3, and describe the pattern.

Week 2 — Patterns and Relationships

Practice

Complete the table for the pattern shown below.

1

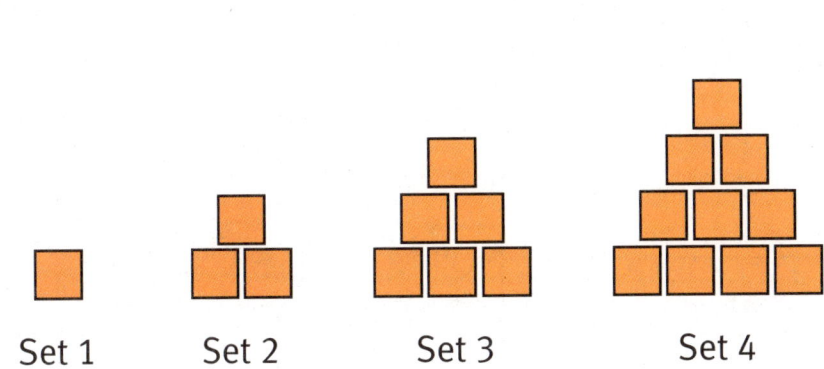

Set	Number of Squares

Complete each input/output table.

2 Candy bars cost 50¢ each.

Input (number of candy bars)	Output (total cost)

3 Jean can ride her bike 20 miles per hour.

Input (number of hours)	Output (number of miles)

Week 3 Patterns and Graphs

Practice

1 The Booster Club sells gourmet cookies for $1.50 each. Complete the input/output table, and graph the pattern.

Input (number of cookies)	Output (total cost)

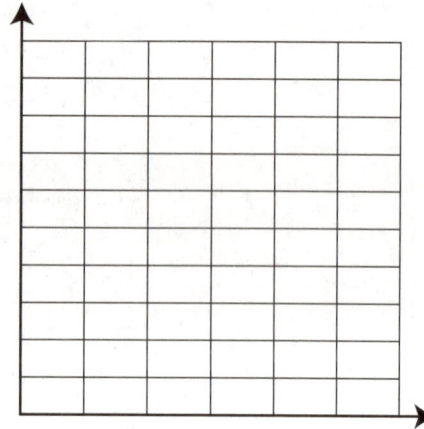

2 Describe the pattern that is shown in the graph.

_____

_____

_____

3 How much would it cost to purchase 10 cookies?

_____

4 What remains the same in the graph?

_____

44 Number Patterns and Relationships • Week 3 Practice

Week 4

Variables and Equality

Practice

1. $x - 5 = 2$

   $x =$ ______

2. $10 - y = 6$

   $y =$ ______

3. $z + 5 = 15$

   $z =$ ______

4. □ + □ + □ = 9

   What is □ ?

5. 22 − △ = △

   What is △ ?

6. ○ + ○ = 12

   What is ○ ?

7. △ + △ = 18

   What is △ ?

8. $m - 6 = 2$

   $m =$ ______

9. Explain how to find the values of the unknowns in the number sentence

   □ + □ + □ = 27

   ____

   ____

   ____

   ____

   ____

Unit 2 Workbook

SRAonline.com

Level F R53180.01